Philip Kollmannsberger

Nonlinear Microrheology of Living Cells

Philip Kollmannsberger

Nonlinear Microrheology of Living Cells

Using Magnetic Tweezers to determine the nonlinear viscoelastic properties of adherent cells

Südwestdeutscher Verlag für Hochschulschriften

Impressum / Imprint
Bibliografische Information der Deutschen Nationalbibliothek: Die Deutsche Nationalbibliothek verzeichnet diese Publikation in der Deutschen Nationalbibliografie; detaillierte bibliografische Daten sind im Internet über http://dnb.d-nb.de abrufbar.
Alle in diesem Buch genannten Marken und Produktnamen unterliegen warenzeichen-, marken- oder patentrechtlichem Schutz bzw. sind Warenzeichen oder eingetragene Warenzeichen der jeweiligen Inhaber. Die Wiedergabe von Marken, Produktnamen, Gebrauchsnamen, Handelsnamen, Warenbezeichnungen u.s.w. in diesem Werk berechtigt auch ohne besondere Kennzeichnung nicht zu der Annahme, dass solche Namen im Sinne der Warenzeichen- und Markenschutzgesetzgebung als frei zu betrachten wären und daher von jedermann benutzt werden dürften.

Bibliographic information published by the Deutsche Nationalbibliothek: The Deutsche Nationalbibliothek lists this publication in the Deutsche Nationalbibliografie; detailed bibliographic data are available in the Internet at http://dnb.d-nb.de.
Any brand names and product names mentioned in this book are subject to trademark, brand or patent protection and are trademarks or registered trademarks of their respective holders. The use of brand names, product names, common names, trade names, product descriptions etc. even without a particular marking in this work is in no way to be construed to mean that such names may be regarded as unrestricted in respect of trademark and brand protection legislation and could thus be used by anyone.

Verlag / Publisher:
Südwestdeutscher Verlag für Hochschulschriften
ist ein Imprint der / is a trademark of
OmniScriptum GmbH & Co. KG
Bahnhofstraße 28, 66111 Saarbrücken, Deutschland / Germany
Email: info@svh-verlag.de

Herstellung: siehe letzte Seite /
Printed at: see last page
ISBN: 978-3-8381-1607-5

Zugl. / Approved by: Erlangen, Friedrich-Alexander-Universität, Dissertation, 2009

Copyright © 2010 OmniScriptum GmbH & Co. KG
Alle Rechte vorbehalten. / All rights reserved. Saarbrücken 2010

Contents

1 **Introduction**   5
   1.1 Preface 6
      1.1.1 Cell mechanics in biology and physics 6
      1.1.2 Mechanical structure of cells 8
      1.1.3 Microrheology of living cells 9
   1.2 Review of cell mechanics 11
      1.2.1 Historical development 11
      1.2.2 Magnetic particle microrheology 11
      1.2.3 Other methods of cell rheology 12
      1.2.4 Summary of current results 12
   1.3 Open questions 14
      1.3.1 Nonlinear microrheology 14
      1.3.2 Stress stiffening or shear softening? 14
      1.3.3 Theoretical description 15

2 **Materials and Methods**   17
   2.1 Hardware 18
      2.1.1 Magnetic Tweezers 18
      2.1.2 Peripheral components 22
      2.1.3 Imaging and data acquisition 23
      2.1.4 Probe particles 25
   2.2 Calibration 26
      2.2.1 Needle sharpening 26
      2.2.2 Hysteresis 26
      2.2.3 Force calibration 28
   2.3 Software 32
      2.3.1 ccd.lib class library 32
      2.3.2 Measurement software 34

  2.3.3 Data analysis in MATLAB 35
 2.4 Procedures 39
  2.4.1 Bead coating 39
  2.4.2 Cell culture 39
  2.4.3 Measurement protocol 39

3 Experimental Results 41

 3.1 Linear creep response 42
  3.1.1 Power law creep 42
  3.1.2 Interpretation of parameters 43
  3.1.3 Statistical analysis 44
  3.1.4 Scaling the creep response 47
  3.1.5 Linearity and superposition 48
  3.1.6 Force dependence of parameters 49
 3.2 Nonlinear differential creep 51
  3.2.1 Differential step protocol 51
  3.2.2 Stress stiffening and fluidization 51
  3.2.3 Prestress determines stress stiffening 54
  3.2.4 Power law exponent and fluidization 54
  3.2.5 Quantification of adhesion strength 57
  3.2.6 Force ramp in the nonlinear regime 58
 3.3 Creep recovery and plasticity 62
  3.3.1 Quantification of creep recoil 62
  3.3.2 Incomplete recovery 64
  3.3.3 Increase of power law exponent 64
  3.3.4 Prestress determines recovery and speed-up 66
  3.3.5 Repeated force steps and preconditioning 68
 3.4 Biological applications 70
  3.4.1 Vinculin as mechanoregulator 70
  3.4.2 FLNa determines active but not passive stiffening 70
  3.4.3 Cancer cell metastasis and cell rheology 73

4 Theoretical Model 75

 4.1 Introduction 76
  4.1.1 Overview 76
  4.1.2 Soft Glassy Rheology (SGR) 77

	4.1.3	The Wormlike Chain (WLC) 80

       4.1.3   The Wormlike Chain (WLC) 80
       4.1.4   Force-induced unbinding of biological bonds 83
       4.1.5   The Sliding Filament (SF) model 85
       4.1.6   Outline for a combined generalized model 86
  4.2 Geometry and stress-strain relationship 89
       4.2.1   Sliding filament geometry 89
       4.2.2   Nonlinear elasticity 89
       4.2.3   Stress-strain curve: numerical results 91
  4.3 Time-dependent behavior 94
       4.3.1   Stress relaxation due to spontaneous unbinding 94
       4.3.2   Force-dependent lifetimes 95
       4.3.3   Numerical results 98

5 Discussion   **103**
  5.1 Limitations and improvements of the experimental setup 104
       5.1.1   Force magnitude 104
       5.1.2   Imaging of intracellular structures 104
       5.1.3   Improvements of the software 108
  5.2 Discussion of the experimental results 111
       5.2.1   Power law rheology 111
       5.2.2   Variation of parameters between cells 112
       5.2.3   Role of prestress 112
       5.2.4   Active or passive stress stiffening? 113
       5.2.5   Quantitative analysis of adhesion strength 114
       5.2.6   Non-recovery, plasticity, and force reversal 114
  5.3 Discussion of the theoretical model 115
       5.3.1   Comparison to the Soft Glassy Rheology (SGR) model 115
       5.3.2   Relation to the „Glassy Wormlike Chain" (GWLC) 116
       5.3.3   Limitations of the numerical results 117
       5.3.4   Uniaxial force generation 117
       5.3.5   Applicability to other systems 117

Appendix   **119**

Bibliography   **120**

1 Introduction

1.1 Preface

In the first part of the introduction, the topic of cell mechanics is motivated by pointing out the important role of the mechanical properties of cells in biology and soft matter physics. A brief overview of the mechanical structure of eukrayotic cells is given. Furthermore, the applicability of rheology to living cells as a means to quantify their material parameters is justified.

1.1.1 Cell mechanics in biology and physics

The cell adopts a unique position as the basic unit of life: every living organism is built of cells, and every individual cell contains the complete genetic information of the organism in its nucleus. The genetic information is encoded in desoxyribonucleic acid (DNA) molecules in the nucleus and is continually translated into thousands of different proteins by the cell, using a complex molecular machinery. Proteins are the building material of the cells and tissues that constitute the structure and physical properties of an organism. These physical properties are an important factor for the ability of the organism to survive and pass on the genetic information in its cells to the following generation. This cycle from genetic information to physical properties and back is illustrated in Fig. 1.1.

Our knowledge about the molecular details of life has grown rapidly in the recent years. This progress is owed largely to the application of quantitative theoretical and experimental methods from physics and chemistry to molecular biology. With the help of these modern techniques, the composition and interaction of proteins can be studied down to the lowest level of structural detail.

In contrast to the fast growing knowledge on the microscopic level, our understanding of the emergent properties of living systems that arise from the cooperative interplay of large numbers of proteins is still limited. The language of systems biology helps to integrate the vast amount of data obtained on the microscopic level into a big picture. It provides a quantitative framework for linking the low-level mechanistic description of biochemical reactions with the high-level functional description of biology.

The systems biological description of cells by means of reaction networks and functional relationships between genes, however powerful it may be, is incomplete as long as the physical and mechanical properties on the integrated level are left aside. Actually, the mechanical properties of cells are essential for understanding many biological processes such as force generation, migration, mechanosensing or morphogenesis.

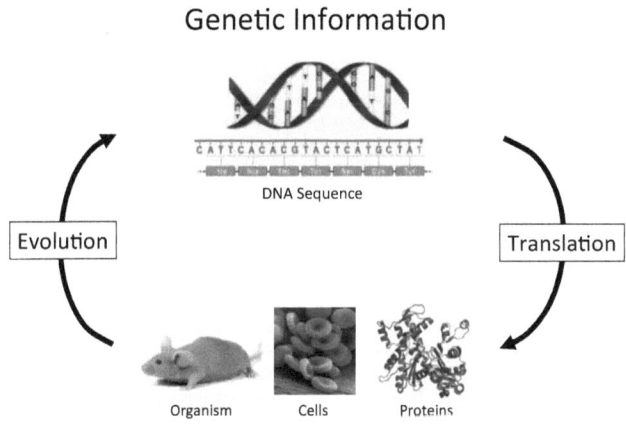

Figure 1.1: The genetic information encoded in the DNA is continually translated into thousands of different proteins using a complex molecular machinery. These proteins build the cells and physical structures of an organism. The mechanical properties of these structures play an important role for the biological function of the organism and the survival probability of the genetic information in its cells during evolution.

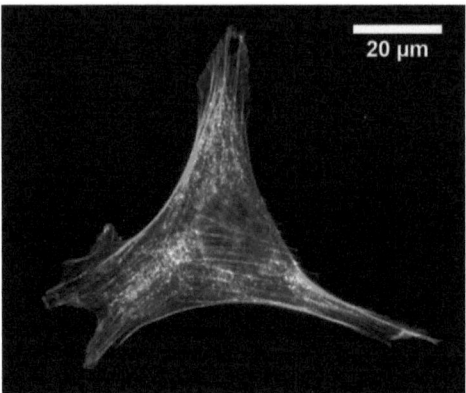

Figure 1.2: Cytoskeleton of a NIH 3T3 fibroblast cell grown on a flat glass surface coated with fibronectin. The filamentous actin cytoskeleton is stained with Alexa-488 phalloidin (cyan). The focal adhesions that link the cell mechanically to the substrate are labelled using paxillin antibodies (magenta). Some of the freely diffusing paxillin proteins in the cytoplasm were also labelled. The activity of myosin motor proteins that contract the actin cytoskeleton is shown by immunolabelling of phosphorylated myosin light chain (yellow). The nucleus of the cell is visible as a dark oval structure in the center. The image was taken using a 100x 1.49 NA objective on a Leica DMI-6000B microscope.

From a biological perspective, cell mechanics should therefore play a central role in any quantitative description of cellular properties. From a physics perspective, the cell as a self-organizing active soft material is by itself a fascinating subject to study.

1.1.2 Mechanical structure of cells

In biology text books, the physical structure of the cell is often simplified as a kind of liquid-filled bubble that serves as a reaction vessel for protein interactions. In reality, however, cells have a highly complex mechanical structure, stabilized by a contractile filamentous soft skeleton that spans the entire cell body (Fig. 1.2). The internal dynamics of the cell as well as the interaction with its environment are strongly influenced by the cytoskeleton. Rather than being a static structure, it undergoes constant remodeling in order to adapt to different external conditions.

1.1 Preface

The mechanical structure of all eukaryotic cells is very similar, therefore general insight can be obtained from mechanical studies of individual representative cell types. The proteins that make up the cytoskeleton of almost all eukaryotic cells are among the best-conserved throughout evolution, which is a hint to their universal role. Actin, the main component of the cytoskeleton, is the most abundant protein in eukaryotic cells and is conserved from yeast to humans [Schmidt 98]. Other filamentous constituents of the cytoskeleton include microtubules – comparably stiff filaments that act as tracks for intracellular traffic – and the more cell-type specific intermediate filaments.

The different types of cytoskeletal filaments are crosslinked by a large number of crosslinking proteins, many of which also play a role in signalling. In addition, the actin network is contractile due to crosslinking with myosin-II motor proteins that actively generate force by converting chemical into mechanical energy. On the molecular level, this mechanism is identical to force generation in skeletal muscle, however the actin cytoskeleton is not as regular as the crystalline structure of muscle filaments.

1.1.3 Microrheology of living cells

While the molecular composition of the constituents of the cytoskeleton remains unchanged, it is their versatile interplay that enables cells to adapt mechanically to a wide variety of demands. Depending on the external environment and the desired biological function, cells can modulate their mechanical state in a surprisingly wide range from almost fluid to almost solid. This ability is essential for carrying out complex mechanical tasks, and derangements play an important role in diseases such as asthma, cardiovascular diseases, or cancer.

The mechanical and flow behavior of soft materials is covered by the science of rheology. Despite the fact that remodeling and response of the cytoskeleton to external stimuli are active biological processes, the mechanical parameters of the cell, such as stiffness and viscosity, can be studied by similar methods of rheology as in non-living soft materials. During the recent years, a wide variety of experimental methods has been developed or adapted for rheological measurements on living cells (Fig. 1.3). The first systematic mechanical studies on biological specimens, however, have already been carried out in the middle of the 19th century. In the following section, the historical development of cell rheology from these early days up to modern precision microrheology is reviewed.

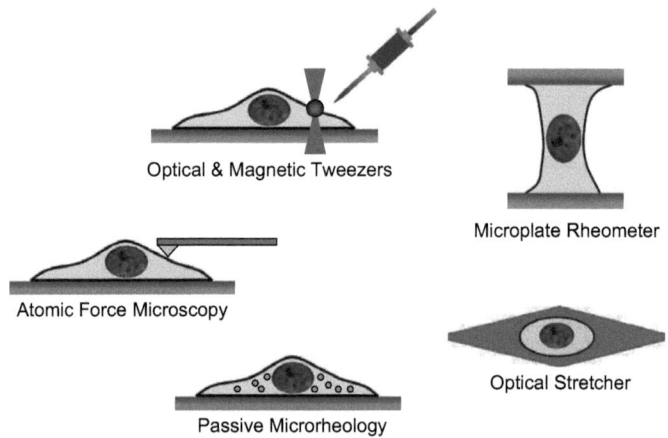

Figure 1.3: Different microrheology techniques have been developed to measure the mechanical properties of living cells. In some of these approaches, the movement of micrometer-sized cytoskeletally bound beads in response to externally applied (Optical & Magnetic Tweezers) or active internal forces (Passive Microrheology) is recorded. Other methods use force transducers (Atomic Force Microscopy, Microplate Rheometer) or trap the whole cell in a dual-beam trap (Optical Stretcher).

1.2 Review of cell mechanics

Starting from the earliest quantitative studies of biorheology in the 19th century, a comprehensive overview of the development and current state of cell mechanics is given in the following section. The different techniques that have been employed to characterize living cells mechanically are introduced. Furthermore, experimental evidence is reviewed for the hypothesis that cells belong to the class of soft glassy materials.

1.2.1 Historical development

Apparently modern concepts often turn out to be surprisingly old: two of the key mechanical characteristics of cells and many biological tissues, namely scale-free stress relaxation and nonlinear stress stiffening, have already been described as early as 150 years ago. In his study of biological tissue elasticity published in 1847, Wertheim [Wertheim 47] showed that tissue stress increases much faster with strain than predicted by Hooke's law. He also describes strain rate dependence and prestress. Even earlier, Wilhelm Weber reported the long-lasting power-law like stress relaxation of spider silk [Weber 35] and undertook first attempts to describe it theoretically [Weber 41]. With the development of linear viscoelasticity, however, the description of stress relaxation by power laws was discarded in favor of mechanistically more intuitive spring-dashpot equivalent circuits with exponential relaxation behavior [Maxwell 67].

1.2.2 Magnetic particle microrheology

The first microrheological studies of the cytoplasm were reported in 1922 by Heilbronn [Heilbronn 22] using an early version of magnetic tweezers. The same method was improved by Francis Crick around 1950, who was later awarded the Nobel prize for the discovery of the spatial structure of DNA. In his studies of the mechanical properties of chick fibroblasts in culture [Crick 50], he laid the groundwork for the development of modern cell mechanics. Despite experimental limitations, he already observed viscoelastic creep and incomplete recovery (Fig. 1.4a) in response to a force step. After being improved further during the second half of the 20th century [Yagi 61, Hiramoto 69, Valberg 87, Guilford 92, Ziemann 94, Fabry 01], magnetic particle microrheology is now a standard technique for studying the mechanical properties of living cells in culture. Micron-sized particles bound to the cell are ideal mechanical probes on a physiologically relevant length scale: focal contacts, which constitute the mechanical window of the cell to its environment, are typically of the same size.

1.2.3 Other methods of cell rheology

Decades after the magnetic particle method had already been established in cell biology, it was discovered that focused laser beams can also be used to exert forces onto small spherical particles [Ashkin 70]. Optical traps allow for much finer control of force and position compared to magnetic tweezers. Their drawback is that the range of applicable forces is much smaller since high laser power would damage living samples. Nevertheless, optical tweezers have been used extensively for cell microrheology [Laurent 02]. An alternative implementation of optical tweezers uses two opposing beams to capture and deform non-adhering cells [Guck 01].

Another new method of microscopic force application found its way from basic physics to biology: it only took five years from the invention of atomic force microscopy (AFM) in 1986 [Binnig 86] to its application in cell biology [Radmacher 92]. Here, force is applied via a flexible force-calibrated cantilever. A similar approach that lies conceptually between that of AFM and a bulk rheometer is the microplate rheometer first introduced by Thoumine and Ott [Thoumine 97]. In this case, cells are allowed to adhere between two plates coated with extracellular matrix protein, one of which is then used to apply and measure stress and strain on the length scale of the whole cell. Instead of a cantilever, a red blood cell aspirated by a micropipette can be used as force transducer [Simson 98].

An approach that is fundamentally different to all previously described methods is passive microrheology. In this case, no external force is applied but the spontaneous thermal or non-thermal motion of a marker particle is used to derive mechanical parameters. The theoretical basis of this method is the Stokes-Einstein relation, which states that in a viscous material with viscosity η at temperature T, the diffusion coefficient D of a spherical particle with radius r depends on the viscosity as $D = k_B T/(6\pi \eta r)$, where k_B denotes the Boltzmann constant. This can be generalized to the fluctuation dissipation theorem, which states for any linear viscoelastic medium that the mean square displacement $\Delta x^2(\omega)$ of a spherical marker particle of radius r is related to the complex shear modulus G^* as $\langle \Delta x^2(\omega) \rangle = k_B T/\pi i \omega G^*(\omega) r$. In cells, the spontaneous motion of intracellular particles has been used to derive a shear and loss modulus [Yamada 00]. A general problem is that the motion of such marker particles in cells is not only due to thermal motion, but also due to active force fluctuations in the cytoskeleton [Raupach 07].

1.2.4 Summary of current results

Despite the diversity of the experimental methods and approaches used in cell microrheology, the consensus result is that in the linear regime of small forces and amplitudes, the creep response $J(t)$, stress relaxation response $F(t)$, and complex shear modulus $\tilde{G}(\omega)$ all follow a power-law in time or frequency, regardless of cell type. This power-law behavior holds over a broad range of frequencies from 0.01 Hz to 1 kHz. In a recent comprehensive study, these findings have been summarized nicely (Fig. 1.4b), and several of the abovementioned methods have been combined to study the contributions of different parts of the cytoskeleton [Hoffman 06].

The power-law viscoelastic response found in cells has inspired the hypothesis that cells belong to the class of soft glassy materials and exist in a metastable state close to a glass transition [Fabry 01]. According to this hypothesis, the rheological power law exponent is a measure for the internal dynamics of the cytoskeleton and can be interpreted as an effective matrix temperature. Subsequently, several studies have been carried out to find further evidence of glassy behavior in cells. The contributions of nonthermal stress fluctuations and the evolution of cell mechanical properties after different waiting times following oscillatory force application have been interpreted in terms of aging and rejuvenation [Bursac 05]. The time course of the development of the shear and loss moduli during recovery from a 25% stretch of the entire cell [Trepat 07] gives further evidence that cell mechanical behavior can indeed be described by Soft Glassy Rheology (SGR).

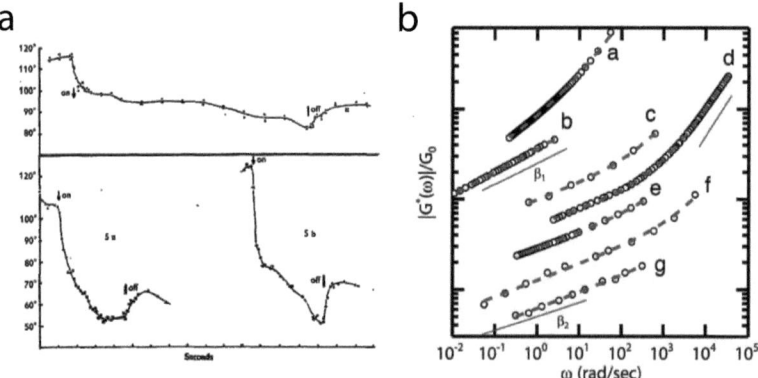

Figure 1.4: a) Creep response of magnetic iron particles in chick fibroblasts, obtained by F. Crick in 1949. From [Crick 50]. b) Summary of literature data on power law rheology of adherent cells. Different active and passive microrheology methods and different cell types were used. From [Hoffman 06].

1.3 Open questions

In the last part of the introduction, open questions in cell mechanics are discussed. The importance of nonlinear microrheology is pointed out, and the seemingly contradicting predictions from linear results regarding the nonlinear behavior are compared. Furthermore, the lack of a comprehensive model of cell mechanics is pointed out.

1.3.1 Nonlinear microrheology

The microrheology studies on cells described in the previous section have in common that they have all been carried out in the linear regime of small forces and deformations. One the one hand, this is due to experimental limitations of the applicable force. On the other hand, there is no standardized protocol to measure and analyze time-dependent rheological properties in the nonlinear regime.

Nevertheless, the high-force nonlinear regime is of physiological relevance, as cells routinely experience large stresses and strains in the body. As a result of the technical limitations of microrheology, an important prediction of SGR concerning the high-force behavior has not been tested directly by experiments so far: if cells were a soft glassy material, they would show shear softening and yielding under large mechanical stresses.

1.3.2 Stress stiffening or shear softening?

There is, however, indirect evidence for a nonlinear behavior that contradicts shear softening as predicted by SGR. Using a combination of magnetic twisting microrheology and traction microscopy, it was discovered that the stiffness of adherent cells is linearly related to their contractile prestress [Wang 01]. Such stress stiffening cannot be explained by SGR, but is reminiscent of the static nonlinear stress-strain relationship observed in most biological materials.

Stretching of whole cells using a microplate rheometer [Fernandez 06] revealed that their static force-length curve is exponential, similar to that of biological tissue on larger length scales [Fung 93]. So far, no systematic time-dependent rheological measurements in the nonlinear regime have been carried out. Such experiments are, however, necessary to probe how soft glassy properties on the one hand and stress stiffening on the other hand contribute to the nonlinear rheology of cells.

1.3.3 Theoretical description

Apart from the inability to account for stress stiffening, the SGR model has another disadvantage: it merely provides a generic phenomenological framework. Although it accurately captures many of the mechanical features of cells, it cannot give any microscopic explanation of the model parameters. In order to lead to new insights, a interpretation of the parameters beyond their simple phenomenological meaning would be necessary.

Other more microscopic models of the cytoskeleton correctly describe nonlinear elasticity on the level of individual filaments and networks [MacKintosh 95, Gardel 04, Storm 05], but do not account for the dynamic and time-dependent mechanical properties of cells. A complete mechanistic model that describes all of the linear and nonlinear mechanical properties of the cell in one comprehensive picture has not been developed so far.

2 Materials and Methods

2.1 Hardware

In the course of this work, a unipolar magnetic tweezers setup was implemented and optimized for high forces. After a short introduction of the basic principle, the different parts of the experimental setup are explained in detail. The choice of the core material and the tip geometry are discussed. Components for image and data acquisition as well as the stage heating needed for live cell microscopy are described. Furthermore, the timing accuracy of the force generated by the magnetic tweezers is characterized.

2.1.1 Magnetic Tweezers

Historical development

The use of magnetic forces in biology dates back to the early 20th century, when Heilbronn, a botanist, inserted magnetic particles into protoplasts and observed their movement in a magnetic gradient [Heilbronn 22]. Similar work was carried out by Freundlich and Seifriz in echinoderm eggs around the same time [Freundlich 23]. In 1950, Crick and Hughes [Crick 50] characterized the viscoelastic response of chick fibroblasts, and later Yagi [Yagi 61] and Hiramoto [Hiramoto 69] further refined the magnetic particle method as a tool to study the viscoelasticity of living cells. Since then, many researchers have utilized and improved different variants of the magnetic particle method as a tool for manipulating biomolecules [Rondelez 05], protein force spectroscopy [Gosse 02], and microrheology studies in soft materials [Ziemann 94] and cells [Bausch 98].

Basic principle

If a particle with volume V and magnetic susceptibility χ is exposed to an external magnetic field H, the field induces in the particle a magnetic moment,

$$m = \chi V H. \tag{2.1.1}$$

If the magnetic field is inhomogenous ($\nabla B \neq 0$), the probe experiences a gradient force,

$$F_x = m \frac{\partial B}{\partial x}, \tag{2.1.2}$$

2.1 Hardware

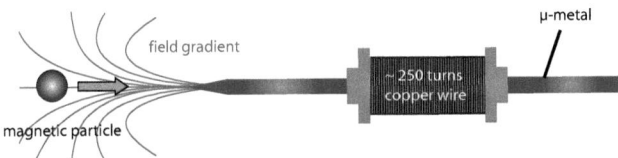

Figure 2.1: Magnetic Tweezers: a high magnetic field gradient builds up around the needle-shaped tip of a superparamagnetic core protruding from a coil with about 250 turns of copper wire. A magnetic particle with positive susceptibility then experiences a force towards the tip.

along the direction x of the magnetic flux. The magnitude of the force F_x depends on the magnetic moment of the probe, which is determined by its size and susceptibility as well as the magnitude of the applied magnetic field, and on the gradient of the field.

This principle can be used for manipulation of magnetic probe particles or to measure the material properties of samples on the micrometer scale. In analogy to their mechanical counterpart, devices that employ this principle are termed *Magnetic Tweezers*.

Geometry

The magnetic tweezers configuration described here was developed with the aim of applying the highest possible force while maintaining precise force control. The magnetic field is generated by a electromagnetic coil, where the field magnitude depends on the current and the number of coil turns. A superparamagnetic core guides the magnetic flux along the axis of the coil and generates a high field gradient around its needle-shaped tip (Figure 2.1).

Magnitude and geometry of the field gradient (and thus force) depend on the shape of the tip: the sharper the tip, the higher the gradient but the faster its decay with distance to the tip [Matthews 04]. The optimum force at distances around 20 μm was observed for a radial symmetric tip with a steep angle of about 30° and a small tip radius of less than 10 μm.

The magnetic needle and solenoid are attached to a micromanipulator at an angle of 45°, with the tip end in the field of view of an inverted microscope. Since the force depends inversely on the distance to the tip, it is desirable to move the end of the needle as close as possible to the probe on the specimen, which requires precise position control and an open chamber design. Both solenoid current and tip position can be manipulated precisely, allowing for accurate force control. In the following paragraphs, the different components of the magnetic tweezers setup are described in more detail.

Core material

The two relevant characteristics for the magnetic core material are its permeability $\mu(H)$, which determines the ratio between magnetic flux density B and given external magnetic field H, and its saturation flux density, B_s. The strength of the gradient depends not only on the shape of the tip as mentioned above, but also on the flux density in the core. Flux density is limited by core saturation as well as solenoid size and current, so both B_s and μ should be high. Moreover, the core material should have low magnetic hysteresis in order to get better control over the applied force.

These requirements are best fulfilled by μ-metal, a superparamagnetic nickel-iron alloy with a high nickel content of about 80%, which is typically used as magnetic shielding material. Several variants from different manufacturers were tested as core material, and the highest forces were obtained with HYMU-80 (Carpenter Technology, Reading, PA) and MUMETALL (Vacuumschmelze GmbH, Hanau).

In Figure 2.4b, the hysteresis curves, from which permeability, saturation and hysteresis can be determined, are shown in comparison. In order to optimize the magnetic properties, the cores should be annealed, i.e. exposed to a high magnetic field at temperatures above 1000°C for several hours followed by a defined cooling protocol. The MUMETALL probe was reannealed after grinding the tip, which explains the lower hysteresivity compared to the HYMU-80 core.

Solenoid

The coil frame is lathed from a brass rod with a 4.5 mm hole in the center to hold the core, and wrapped with about 250 turns of 0.5 mm lacquered copper wire. The wire is connected to a coaxial cable which runs to the power supply. The core is fixed by two screws at either end of the coil body. An additional envelope for water cooling can be put over the discs. Brass is chosen over the less noble aluminum for the coil body

2.1 Hardware

Figure 2.2: Photograph of the setup. An inverted microscope with the Magnetic Tweezers attached to a micromanipulator stands on top of an air table, with the current source on the right and the camera controller on the left side. Experiments are controlled from the PC left of the microscope, which is equipped with data acquisition (DAQ) and framegrabber boards.

since the latter would likely be corroded by the nickel-alloy core. The core itself has a diameter of 4.5 mm and a length of 100 mm.

2.1.2 Peripheral components

Current source

The current source for the solenoid current is custom built and can generate a current of up to 3 A at an output power of 35 W. It consists of an OPA-549 high current operational amplifier (Texas Instruments, Dallas, TX) in a 1:1 voltage-to-current conversion circuit.

The operational amplifier was chosen for high bandwidth and low noise and drift. It is powered by a regulated power supply which was either custom-built or commercial (IHCC15-3, International Power, Oxnard, CA). The actual current through the solenoid is monitored by picking off the voltage over the 1 Ω feedback resistor which runs in series to it. The input of the current source was connected to the analog output of the data acquisition board in the PC.

Hall probe

The magnetic flux density of the microneedle is monitored by a miniature Hall probe (SS496A, Honeywell, Morristown, NJ) attached to the rear end of the core. Its main purpose is for calibration of magnetic hysteresis, but in principle the magnetic flux signal could be used for real-time force regulation as well. The sensor cable is connected to a small box housing a 9 V battery to power the sensor and a voltage divider circuit with a potentiometer to control the offset of the Hall voltage.

Micromanipulator

The microneedle is attached to a micromanipulator (InjectMan NI-2, Eppendorf AG, Hamburg, Germany) for precise positioning of the tip with an accuracy of 40 nm. The micromanipulator has a RS-232 interface and can be controlled remotely from a personal computer, or manually using a joystick. A custom adapter was built to mount the needle to the micromanipulator at the proper distance to get the needle into the field of view of the microscope.

Stage heating

In order to provide a proper physiological environment for cell biological experiments, the temperature of the sample should be close to 37°C. This was accomplished by attaching a network of power transistors connected to a current source to the lower side of the sample stage. The waste heat of the transistors heats the stage, and an analog PID control loop in the current source maintains a constant stage temperature.

For the motorized stage of the Leica DMI-6000B, an alternative heating system using InZnO-coated glass bottom wells (Delta T, Bioptechs) was developed. Here, the heating current of the Delta T dishes is controlled by a digital PID controller implemented on an ATMEGA-168 microcontroller.

2.1.3 Imaging and data acquisition

Imaging system

The magnetic tweezers setup is built around an inverted microscope (Leica DM-IL or Leica DMI-6000B) with a 40x 0.6 NA non-magnetic objective. The microscope stands on top of an anti-vibration air table in order to mechanically uncouple it from the environment. An overview of the setup is shown in Figure 2.2. For accurate tracking of magnetic beads, bright-field illumination is used. Images are taken by a 12 bit black-and-white CCD camera (Hamamatsu Orca-ER) with a resolution of 1344 x 1024 pixels. The camera was chosen for the low electron noise and the anti-blooming feature of the built-in CCD chip (Sony ICX285). The frame rate is typically between 8 fps for full frames and 40 fps for subframes with 1344 x 128 pixels. The digital images are sent to the PC for processing either through a framegrabber card (Coreco PC-DIG) or directly via the firewire interface of the camera. The shutter of the camera is triggered remotely from the PC. The data flow of the setup is shown in Figure 2.3.

Data acquisition hardware

The voltage signal for controlling solenoid current and the image acquisition trigger signal are generated by a PC equipped with a 16 bit data acquisition (DAQ) board (NI-6052E, National Instruments, Austin, TX). For every captured frame, solenoid current and magnetic flux density are simultaneously recorded using the analog-to-digital AD converter of the NI-6052E board. The CCD camera is triggered using the card's timer output.

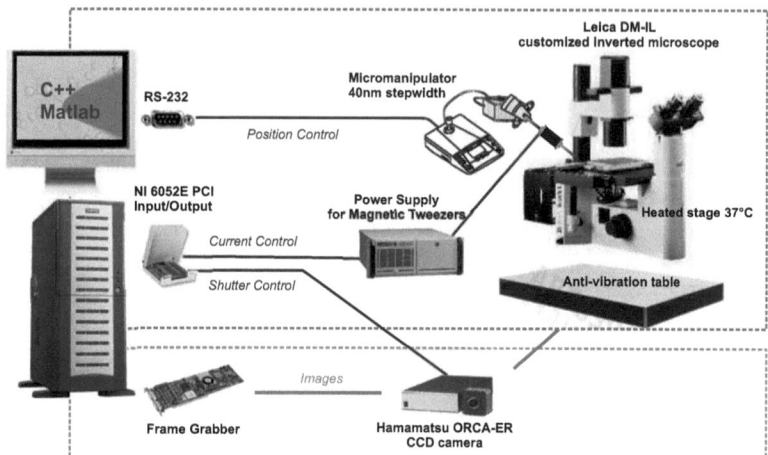

Figure 2.3: Schematic of the setup. The Magnetic Tweezers is attached to a remote-controlled micromanipulator mounted on a Leica customized inverted microscope with a heated stage. The microscope is mechanically isolated with an anti-vibration air table. Images are taken with a CCD camera and transferred to a PC using a framegrabber board. Camera shutter and magnet current are connected to the output channels of the NI DAQ board and controlled from the PC using custom software written in C++.

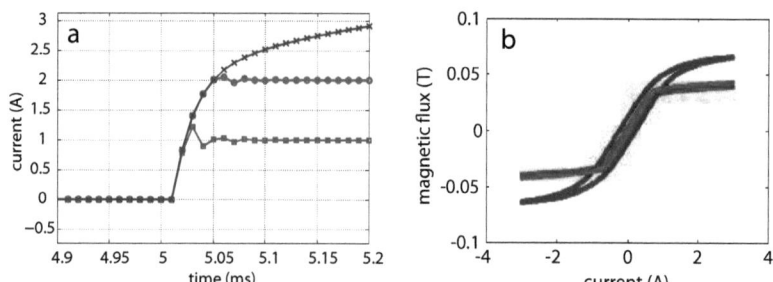

Figure 2.4: (a) Current is switched on at t = 5 s. Rise time of the current and hence force is about 200 μs for a current of 3 A. (b) Magnetic flux vs. coil current for HYMU80 (blue) and MUMETALL (red and green). Hysteresivity and saturation are higher in HYMU80 compared to MUMETALL.

2.1 Hardware

	Bangs Labs Compel Beads	Spherotech CFM-40-10	Invitrogen Dynabeads 450	Custom Fe_3O_4
diameter	5.8 μm	4.4 μm	4.5 μm	~4.5 μm
F at 3 A, 20 μm	2 nN	3.5 nN	12 nN	60 nN
F/V	0.02 nN/μm^3	0.08 nN/μm^3	0.25 nN/μm^3	1.25 nN/μm^3

Table 2.1: Comparison of different types of beads in terms of force-to-volume ratio. Force is given at a driving current of 3 A and a bead-to-needle distance of 20 μm.

Image acquisition and driving current are synchronized, and a hardware generated transistor-transistor logic (TTL) pulse is used to define the start of an experiment. To ensure accurate timing and a fast onset of force despite the high inductivity of the coil, the actual coil current was monitored by measuring the voltage across a 1 Ω resistor in series to the solenoid. The rise time of the current and hence force was in the order of 200 μs (Figure 2.4a).

2.1.4 Probe particles

As magnetic probes, different types of spherical magnetic beads with a radius between 0.5 μm and 9 μm were used. Such beads are commercially available as part of immunolabeling assays. While superparamagnetic beads are only temporarily magnetizable since their grain size is too small for magnetic domains to develop, ferromagnetic beads keep their magnetization after the external field is switched off.

It is desirable to have a narrow distribution of bead sizes in order to get a reliable force calibration. To get high forces, high susceptibility of the beads is required. We characterized different types of magnetic beads in terms of their monodispersity and susceptibility (see table 2.1). Of all commercially available paramagnetic beads that were tested, best results were obtained with Dynabeads M-450 (Invitrogen, Carlsbad, CA).

The only beads that showed even higher susceptibility were custom made ferromagnetic Fe_3O_4 beads and commercial polydisperse carbonyl iron spheres (BASF, Ludwigshafen, Germany). Details on surface functionalization are given in section 2.4.1.

2.2 Calibration

To turn a sharp nail into a precision instrument for microrheology, it is essential to have accurate control over the applied force. This requires knowledge about how the force is influenced by the distance between probe and tip, by the coil current or external field, and by magnetic hysteresis. For this reason, a calibration procedure was developed that yields a list of altogether ten calibration parameters which fully characterize these dependencies. In this section, the different steps of the calibration procedure are described.

2.2.1 Needle sharpening

Since the sharpness of the needle tip is essential for generating high gradients and forces, the needle was resharpened before every new calibration. Accordingly, the calibration had to be redone after every modification of the needle shape and the corresponding gradient field.

Coarse sharpening was done using a grinding machine. After that, superfine abrasive paper (#4000, Struers GmbH, Willich) was put flat on a table and the needle tip was slided carefully back and forth while rotating and pulling it slowly towards the edge. The shape was repeatedly checked under the microscope after cleaning it with an ethanol-drenched cotton bud to remove metal particles.

Variations of needle shape revealed the best force characteristics (highest gradient in front of the tip) for a rather blunt tip angle of about 30° measured from the core axis, and a very sharp tip end with a radius smaller than 10 μm.

2.2.2 Hysteresis

Degaussing

The magnetic needle showed hysteresis, which means that permanent magnetic domains persist within the material and magnetic flux does not completely vanish after switching off the external field. The hysteresis can worsen if the needle is heated up as a result of machining, such as lathing or grinding. One way to overcome this is by annealing the needle material at high temperatures of over 800 °C in a strong magnetic field, which restores the superparamagnetic properties.

The MUMETALL core was sent back to the manufacturer for reannealing after machining, while the HYMU-80 core remained unannealed, resulting in different areas of the

2.2 Calibration

hysteresis loops in Figure 2.4b. Prior to every measurement, the remanent magnetization of the unannealed core was erased by a 50 Hz sinusoidal degaussing pulse with a duration of 3 s and a linearly decaying envelope from 3 A to 0 A, which creates a random orientation of the magnetic domains in the material and, hence, zero total flux.

Anti-voltage

Even after degaussing the core, there was a non-vanishing magnetic gradient possibly caused by distortion of the earth magnetic field around the tip. Consequently, magnetic beads in front of the needle still experienced a non-vanishing force and moved towards the tip. To cancel this remaining force, an additional anti-voltage was applied after degaussing.

In order to find the correct anti-voltage, beads immersed into a low-viscosity (350 mPa·s) silicon oil were moved close to the needle tip, and different voltages were applied after degaussing the core. The value of the anti-voltage was then determined as the voltage were the least movement of beads close to the tip was observable. The Hall probe was attached to the proximal blunt end of the needle to monitor the magnetic flux through the core, and the offset of the output signal was set to zero to define the zero point of flux as the zero point of force.

Coercitive current

To reset the force to zero during an experiment without disturbing the specimen by the degaussing pulse, the remanent magnetic field was cancelled by superimposing a compensatory magnetic field generated by the solenoid. The magnitude of the required coercitive current depends on the magnetic history of the material, i.e., the highest magnetic flux it has been subjected to.

This compensation current was calibrated in the following way. After application of a brief 1 s magnetization current, a compensation current is applied. The resulting magnetic flux is recorded for different combinations of magnetization and compensation currents. The compensation current necessary to generate zero flux is determined for every magnetization current, and the relation between different magnetization currents and their respective compensation currents is fit to a fourth-order polynomial (Figure 2.6d), resulting in five parameters to describe hysteresis compensation.

This force-based procedure to determine the coercitive current is superior to using the hysteresis loop itself, since it ensures that the actual force is set to zero even after the core has been magnetized by an arbitrary current. These calibration procedures for anti-voltage after degaussing and hysteresis compensation without degaussing was repeated for each core-coil combination.

2.2.3 Force calibration

Force is calibrated by measuring the velocity of a bead moving through a viscous fluid – a commonly used method for measuring microscopic forces. It is simplified by the negligible inertia of the small particles and by their spherical shape. The viscous drag F onto a spherical particle with radius r moving with a (laminar) velocity v is described by Stokes' law,

$$F = 6\pi\eta r v, \qquad (2.2.1)$$

where η denotes the viscosity of the background medium.

A small number of the beads to be calibrated are diluted in uncured PDMS (Polydimethylsiloxane, Sigma-Aldrich, St. Louis, MO) with a calibrated viscosity between 0.1 and 10 Pa·s. The needle is then immersed into the solution, and bead movements are tracked during repeated current on-off cycles, with each on or off phase lasting for 1 s. Settling of the viscous fluid causes a constant bead drift, which is determined during the current-off phases. The drift-corrected velocity is then used to calculate the force acting on the bead.

Force-distance curves for multiple beads and currents are recorded (Figure 2.6a), and a simple mathematical expression is fitted to the data of all beads, distances, forces, and currents:

$$F(d, I) = F_0 \left(\frac{d}{d_0}\right)^{c(I)}. \qquad (2.2.2)$$

This equation describes the relationship between force F, current I, and distance d, using two scaling parameters F_0 and d_0 for force and distance, and a current-dependent distance exponent, c, describing the slope of the $F - d$ curve in a log-log plot (Figure 2.6b). For instance, $c = -2$ corresponds to $F \propto 1/d^2$ and $c = -1$ to $F \propto 1/d$.

This current-dependent exponent is fitted to the Fermi-like function,

$$c(I) = \frac{c_1}{1 + c_2 \exp(c_3 I)}, \qquad (2.2.3)$$

as shown in (Figure 2.6c), using three parameters $c_{1...3}$. With a total of five parameters, this calibration formula accurately describes the force at any measured distance and

2.2 Calibration

current in a phenomenological way.

Equation (2.2.2) can be rearranged to compute the driving current that is necessary to obtain a desired force at any given distance between the bead and the needle,

$$I(F,d) = \frac{1}{c_3} \ln \left\{ \frac{1}{c_2} \left(c_1 \frac{\ln(d/d_0)}{\ln(F/F_0)} - 1 \right) \right\}. \quad (2.2.4)$$

Alternatively, Equation (2.2.2) can be rearranged to compute the needle position, i.e., the bead-needle distance that is necessary to obtain a desired force at a given current:

$$d(F,I) = d_0 \left(\frac{F}{F_0} \right)^{\frac{1}{c(I)}}. \quad (2.2.5)$$

The parametrized relation between force, current and distance allows for real time force control during experiments. Needle position and/or coil current are adjusted after every captured frame according to the observed bead movement and desired force. The implementation details of this force feedback mechanism are described in section 2.3.2.

Angle dependence

In general, the force vector was found to be perpendicular to the nearest surface of the core. For beads within an angle of about 120° around the tip and a distance below 100 μm from the tip, the force vector points towards the center of the circle describing the tip. For these beads, the magnitude of force depends only on the Euclidean distance to the tip, and is independent of the bead position relative to the needle axis (Figure 2.5).

This property of the force-distance relationship greatly simplifies the force calibration procedure and magnetic tweezer experiments. For instance, without loss of accuracy, the forcing direction and the needle axis need not be precisely aligned, and the needle tip can be moved a few micrometers above the z-plane of the bead to avoid contact between the needle and the substrate to which the bead adheres.

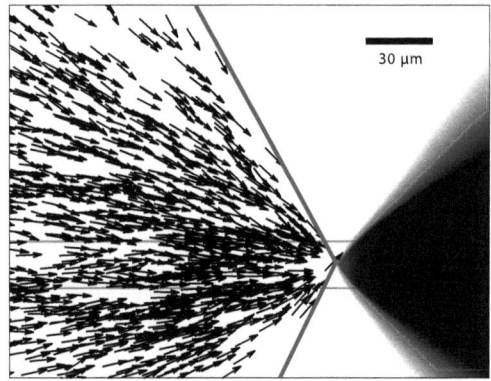

Figure 2.5: The magnitude of force within an angle of about 120° around the tip depends only on the Euclidean distance to the tip. The force vectors point towards the tip.

2.2 Calibration

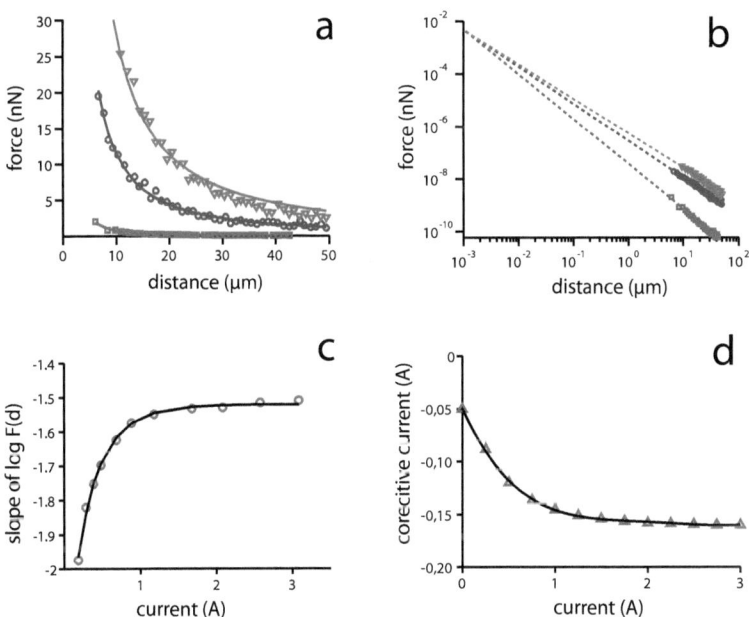

Figure 2.6: Force-distance data (dots) and fit (lines) in linear (a) and double-logarithmic (b) representation, and dependence of the slopes on current (c). Hysteresis compensation currents and the polynomial fit curve are shown in (d).

2.3 Software

The various hardware components are linked to and coordinated from a standard personal computer (PC). In the following section, the computer software developed for data acquisition and data analysis is described. Data acquisition software consists of a C++ class library for hardware control and image processing, and two application programs built on top of it. For data analysis, a suite of user-friendly graphical tools was developed under Matlab.

2.3.1 ccd.lib class library

Concept

Data acquisition software is based on a custom C++ library (ccd.lib) initially developed by J. Pauli [Pauli 05] and further improved in the course of this work. This library contains modules for controlling the different hardware components (camera, current source, micromanipulator, microscope etc.) as well as image processing algorithms and memory management. The concept of a class library has several advantages compared to standalone programs:

- Multiple applications can be developed without rewriting the hardware part.
- Several processes running at the same time can share the same frame buffer.
- Improvements of the library can be propagated to all application programs that use it without changing their source code.

C++ was chosen as a programming language since it offers the possibility of modular object-oriented design while maintaining full compatibility with the low-level C routines supplied by the hardware manufacturers.

Hardware control

The camera used (Hamamatsu Orca-ER) comes with a development kit including a static C library (dcamapi.lib), which provides functions for controlling image capture and timing. These hardware-specific functions are wrapped in general classes for CCD camera control to make the application programs independent of the specific camera model used.

2.3 Software

The same was done with the micromanipulator (Eppendorf InjectMan NI2) which is accessed over the serial interface. The commands specific to this model were encapsulated into a generic micromanipulator class. The other hardware modules (e.g. microscope, focus drive, shutter) of the class library are not described here since they are not used for the magnetic tweezers setup.

Most parts of the library are platform independent, only the parts relying on externally supplied hardware drivers and the implementation of the frame buffer are specific to the Windows operating system. Development was done in the Visual .NET programming environment using the included C++ compiler.

Bead tracking

Bead positions are determined by pattern matching and then tracked through subsequent frames using an intensity-weighted center-of-mass algorithm. The corresponding functions are called findBeads and trackBeads.

In order to find transparent beads, which appear as a dark ring with a bright spot in the center in transmitted light illumination, all bright spots above a certain threshold are checked for a dark surrounding whithin a defined radial interval. The ratio between the intensity in the dark part and in the center and background is set by two threshold values (HighThreshold and LowThreshold), and the outer radius by the parameter BeadRadius. Non-transparent beads do not have a bright central spot, they are found by identifying circular dark regions of the respective size.

In case the criteria for a bead of either type are fulfilled, the actual position is determined with subpixel accuracy by finding the intensity center-of-mass (CoM) within a quadratic window around the center with size set by the parameter BeadWindow. This centering is iterated until the position does not change any more between two iterations. The parameter BeadDistance sets the minimum allowed distance between two beads, one of which is discarded otherwise.

The accuracy of the centering algorithm depends on the intensity resolution and noise of the CCD camera, as well as the magnification and resolution of the microscope objective. In order to quantify the position detection of the measurement system, the position of several beads immobilized on a coverslip was tracked over 100 frames, and the correlated motion of all beads was subtracted from the position signal of each single bead. This yielded a position signal with a SD of less than 1.5 nm for a 40x 0.6 NA objective using the Hamamatsu Orca ER camera.

Needle tracking

The position of the needle tip in the field of view is determined by erosion and dilation operations on a downscaled image, followed by thresholding and segmentation. This yields a binary image of needle and background that is then used to find the needle position. The Euclidean distance of any point in the field of view to the nearest point on the needle is stored in a distance matrix where it can be looked up quickly during the time-critical phase of an experiment.

2.3.2 Measurement software

Tweezers.exe **for live control**

This small graphical user interface tool was developed to quickly set the coil current and degauss the magnet. No additional functionality such as image acquisition is provided. It is used, e.g., during calibration for determining the anti-voltage (see section 2.2 on calibration), where the core is repeatedly degaussed, different anti-voltages are set and the remaining bead movement is monitored under the microscope. Another possible application are cell biological experiments where image acquisition is run independently using other programs.

The live control tool is implemented as a single-window application in C++ using the MFC (Microsoft Foundation Classes) toolkit. The only external library is nidaqmx.lib which is needed to access the analog output channel of the DAQ board. Due to the multithreading capability of this library, simultaneous monitoring of the input channels using other programs such as the *Measurement and Automation Explorer* supplied with the DAQ hardware is possible, e.g., to monitor the Hall probe current during calibration.

Tweezers_CL.exe **for measurement**

During a microrheological experiment, the measurement software needs to carry out several tasks simultaneously: synchronized image acquisition and force control, particle tracking, updating the magnet position, as well as displaying and saving images. For this purpose, a C++ command line application was developed that uses multithreading to coordinate these tasks.

While a command line application is easier to maintain than a graphical user interface when using multiple threads, it does not allow for graphical output. Live image display is therefore done by a separete application that uses the same frame buffer memory for

2.3 Software

image data. An alternative approach using the LabVIEW platform was discarded despite the superior graphical possibilities, since it does not offer the same flexibility and timing reliability as custom C++ code.

Camera control, image processing and frame buffers as well as an interface for micromanipulator remote control are implemented in the ccd.lib class library. Several low-level C libraries provided by manufacturers or freely available are used to access the hardware:

- nidaqmx.lib, provided by National Instruments, to access the analog and digital input/ouput channels of the DAQ hardware.

- dcamapi.lib, provided by Hamamatsu, for setting camera parameters and acquiring images.

- ctb.lib, freely available from IFTOOLS GbR, for easy platform-independent access to serial ports, used for controlling the micromanipulator.

The type of experiment to be carried out and the necessary parameters are passed to the program using a configuration text file with a defined syntax. The functionality for reading in the parameters is provided by the freely available ConfigParser library.

Once started, the program switches between an interactive mode where experiment parameters can be edited while images are acquired and displayed asynchronously, and the time-critical measurement mode with limited user interaction. During measurements, the preset force is maintained by continuously updating the solenoid current or by moving the solenoid according to the realtime bead positions such that the needle-tip to bead distance is kept constant (Figure 2.7). Bead and needle coordinates as well as the monitored current are saved in ASCII format.

2.3.3 Data analysis in Matlab

MT_Browser

The data are analyzed and vizualised under MATLAB using different custom-written graphical user interfaces (Figure 2.8). In the first stage, configuration parameters and raw data from the experiment are collected from the various ASCII files and converted to a common data format described in the Appendix. The corresponding tool, called MT_Browser.m, is common to all experimental protocols, whereas subsequent stages for rheological analysis are necessarily specific to the applied force protocol.

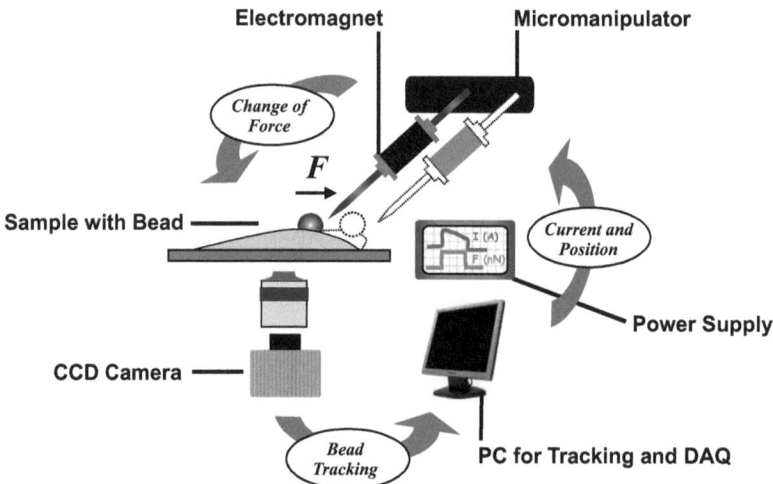

Figure 2.7: Schematic of the control loop that keeps the preset force: bead positions are tracked in realtime, and current or needle position are updated accordingly.

2.3 Software

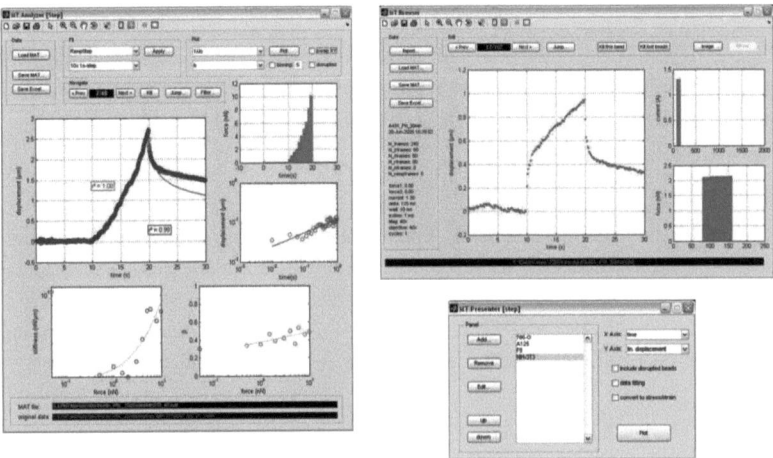

Figure 2.8: Graphical tools for converting (left), analyzing (top right) and displaying (bottom right) the results of microrheological experiments were developed under MATLAB.

Different measurements stored in different locations can be merged. Displacement and applied force over time are displayed for every bead, together with parameters of the experiment. Images or, if present, movies of the measurement can be viewed. Based on this information, the user can selectively delete erroneous data (e.g., beads that were not bound to a cell) to ensure successful fitting of the data in subsequent steps.

After viewing, the raw data can either be stored in a binary .mat file for the later stages of analysis, or exported to a MS Excel spreadsheet.

MT_Analyzer

In the second stage of analysis, various models can be fit to the experimental data, and the resulting parameters are displayed. The tools used in this stage are specific to the applied force protocol. Currently, three different force protocols can be analyzed: one or more repeated constant force steps (*Creep*), an increasing staircase-like force (*Step*), or a linear increasing force (*Ramp*).

The user interface is similar in all three cases. Displacement, force, and various fit parameters are displayed, and different models for the fit can be chosen using drop-down menus. Filters can be applied to remove drift and high-frequency noise from the data or to remove disrupted or erroneous beads. Individual or averaged fit parameters and their distributions can be plotted in various ways. After fitting, the data can be stored in binary format including the fit parameters, or exported to a MS Excel spreadsheet.

MT_Presenter

After fitting the data, the results of different experiments can be presented and compared graphically. The tools for this third stage are again specific to the experimental protocol. Binary data files can be loaded, renamed, and their contents displayed in a variety of ways. The resulting figures can be used directly for publication, e.g. by using the PostScript export feature of MATLAB.

2.4 Procedures

2.4.1 Bead coating

For experiments on cells, beads are coated with different ligands which bind selectively to different membrane receptors. For microrheological measurements, extracellular matrix proteins like fibronectin or collagen are used which bind to transmembrane integrin receptors. Upon ligand binding, integrins are activated and form clusters. As a result, focal adhesion proteins aggregate on the intracellular side of these clusters and link the integrin to the actin network, leading to a firm mechanical connection between the bead and the cytoskeleton (Figure 2.9b).

Epoxylated 4.5 μm beads were washed in PBS and incubated with human fibronectin (100 mg/ml, Roche Diagnostics, Mannheim, Germany) in PBS at 4°C for 24 h. After incubation, the beads were washed three times in PBS and stored at 4°C for experiments. Fibronectin (FN) contains a RGD binding motif that is specifically recognized by integrins $\alpha_v\beta_3$ and $\alpha_5\beta_1$, so synthetic RGD peptide can be used as well instead of full-length FN.

2.4.2 Cell culture

Cells were maintained in low-glucose (1 g/L) Dulbecco's modified Eagle's medium supplemented with 10% fetal calf serum (low endotoxin), 2 mM L-glutamine, and 100 U/ml penicillin streptomycin (DMEM complete medium, all from Biochrom, Berlin, Germany). Eighty percent confluent, adherent cells were detached using Accutase (PAA Laboratories, Linz, Austria), seeded at a density of 10^5 cells onto 35 mm culture dishes (Nunclon Surface, Nunc, Wiesbaden, Germany) in DMEM complete medium and incubated at 37°C and 5% CO_2 overnight.

2.4.3 Measurement protocol

Before measurements, the beads were sonicated and added to the cells (10^4 beads/dish) and incubated in 5% CO_2 at 37°C. After 30-60 minutes of incubation, most beads were tightly bound to the outside of cells (Figure 2.9a), whereas after 24 hours, beads were typically taken up (phagocytozed) by the cells. In order to get reproducible results and to ensure proper bead binding, every dish was incubated with beads for 30 minutes, and subsequent measurements took no longer than 30 minutes, resulting in 20-30 measurements per dish.

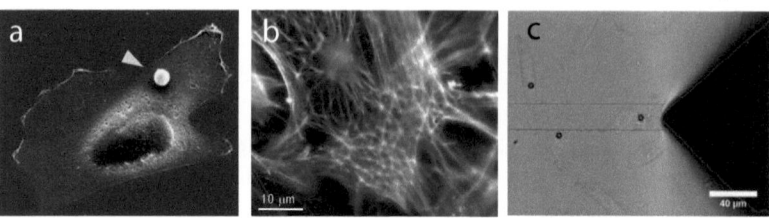

Figure 2.9: (a) REM image of a 4.5 μm superparamagnetic bead (arrow) bound to a fibroblast cell. (b) Fluorescently labelled actin cytoskeleton of a human airway smooth muscle cell with embedded bead (from [Fabry 01]). (c) Bright field image of beads bound to living epithelial cells during a measurement. The tip of the Magnetic Tweezers can be seen on the right.

Before starting a measurement, the needle tip is placed at a distance between 20 and 30 μm from a bead bound to a cell (Figure 2.9c). To ensure that cells had not experienced any significant forces resulting from a previous measurement, the needle was moved at least 0.5 mm between any two measurements.

3 Experimental Results

3.1 Linear creep response

In this chapter, the creep response of various cell types to small applied forces is measured. Power law time dependence, heterogeneity and linearity of the creep response are determined and compared to results from earlier studies using other methods. The validity of the scaling relation that was recently discovered in the frequency domain [Fabry 01] is assessed in the time domain. Higher forces result in a force dependence of parameters while the power law time dependence remains valid.

3.1.1 Power law creep

The creep response of beads bound to adherent cells was determined by applying a constant force F_0 during a period of 10 seconds and recording the resulting bead displacement, $d(t)$. The first frame was taken 10 ms after the onset of force; subsequent frames were recorded at a rate of 40 Hz or every 25 ms, yielding bead position data for times between 0.01 s and 10 s (Figure 3.1). Individual beads moved over time according to a weak power law,

$$d(t) = d_0 \left(\frac{t}{t_0}\right)^\beta, \qquad (3.1.1)$$

with typical exponents β between 0.1 and 0.4 (Figure 3.2). In this formula, time is arbitrarily normalized to $t_0 = 1$ s to get a dimensionless exponent. The choice of t_0 merely affects the definition of $d_0 = d(t = t_0)$.

In order to facilitate comparison with other cell mechanical experiments, the creep compliance $J(t)$ was determined from the bead displacement $d(t)$ as follows. Stress σ_0 is estimated as the applied constant force F_0 divided by the effective area A, approximated by the bead cross section, $A = r^2\pi$. Local strain $\gamma(t)$ is defined as the total displacement divided by the bead radius r, the characteristic length scale of the experiment: $\gamma(t) = d(t)/r$. The resulting time dependent creep compliance is proportional to the measured displacement:

$$J(t) = \frac{\gamma(t)}{\sigma_0} = \frac{d(t)}{r} \cdot \frac{r^2\pi}{F_0} = \frac{r\pi}{F_0} \cdot d(t). \qquad (3.1.2)$$

Hence,

$$J(t) = J_0 \left(\frac{t}{t_0}\right)^\beta \qquad (3.1.3)$$

3.1 Linear creep response

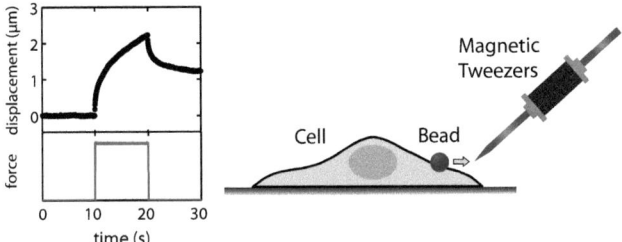

Figure 3.1: The gradient force generated by Magnetic Tweezers acts on superparamagnetic beads coated with fibronectin which are bound to the cytoskeleton of adherent cells via integrin receptors. The creep response was determined by applying a constant force during a period of 10 seconds and recording the resulting bead displacement.

with

$$J_0 = \frac{r\pi}{F_0} \cdot d_0. \qquad (3.1.4)$$

Displacement and creep compliance are related by a constant factor, $r\pi/F_0$. The power law exponent is not affected by the conversion from displacement to creep compliance. Typical displacements at $t = 1$ s and $F_0 = 1$ nN are in the order of 1 μm, yielding a creep compliance of $J_0 \approx 7 \times 10^{-3}$ Pa^{-1}.

3.1.2 Interpretation of parameters

The weak power law relationship between time and displacement remained valid for all cell types, pharmaceutical interventions and genetic modifications that were tested in the course of this work. J_0 varied between 10^{-3} Pa^{-1} and 5×10^{-2} Pa^{-1}, and the power law exponent was between 0.1 and 0.5. The magnitudes of stiffness or shear modulus and power law exponent are consistent with earlier measurements of the linear rheology of adherent cells using other methods [Hoffman 06, Fabry 01].

As a result of the power law fit, two parameters are sufficient to describe the creep response of cells over at least four orders of magnitude of time. This is in contrast to standard viscoelastic analysis, where a linear combination of discrete relaxation modes is used to fit the creep response. Fitting our data to that approach would yield at least two parameters per order of magnitude of time, corresponding to spring constants and

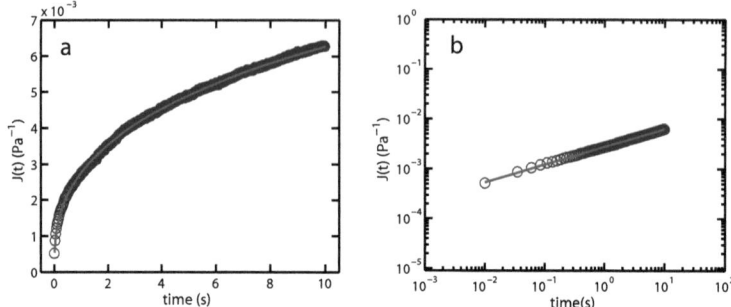

Figure 3.2: Representative creep response of a bead bound to a NIH 3T3 fibroblast in linear (a) and log-log representation (b). The bead displacement in response to a force step is related to the creep response $J(t)$ in units Pa^{-1} by a constant geometric factor. Solid lines: fit to the power law relation eq. (3.1.3).

damping coefficients of the respective spring and dashpot elements. Given the complex and heterogenous structure of the cell, it seems questionable that a small number of discrete springs and dashpots would be the adequate approach to describe its mechanical properties.

The tradeoff of the smaller number of fit parameters of the power law fit is a more difficult interpretation of the parameters, which do not represent spring-dashpot elements as in standard viscoelasticity. The inverse of the prefactor J_0 corresponds to a shear modulus G_0 at a fixed time and is therefore a measure for stiffness. As illustrated in Figure 3.3, the power law exponent can be interpreted as the viscoelastic state of the cell between a purely elastic solid ($\beta = 0$) and a purely viscous fluid ($\beta = 1$). These two parameters can be used to quantify differences in the mechanical properties between different cell types, different bead coatings or binding times, or due to pharmaceutical or biochemical interventions.

3.1.3 Statistical analysis

The exponent β is normally distributed between different cells of the same type, while the creep compliance J_0 follows a log-normal distribution (see Figure 3.4), both consistent with literature [Desprat 05, Fabry 03]. The origin of the log-normal distribution remains unclear – it is not a result of the bead-cell geometry since the same distributions

3.1 Linear creep response

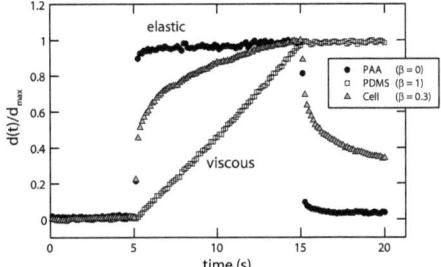

Figure 3.3: The power law exponent β can be interpreted as the viscoelastic state between a purely elastic solid ($\beta = 0$) and a purely viscous fluid ($\beta = 1$). This plot shows the creep response of a bead bound to a cell in comparison to beads in a linear elastic gel (polyacrylamide, PAA) and in a viscous calibration standard (polydimethylsiloxane, PDMS).

can be observed in whole-cell stretching experiments [Desprat 05].

The type of distribution determines how the mean and standard error of a data sample have to be calculated. For n measurements of the normally distributed exponent β, the mean is simply the arithmetic mean of all measurements,

$$\overline{\beta} = \frac{1}{n} \sum_{i=1}^{n} \beta_i, \qquad (3.1.5)$$

with standard error

$$SE_{\overline{\beta}} = \sqrt{\frac{1}{N(N-1)} \sum_{i=1}^{n} (\beta_i - \overline{\beta})^2}. \qquad (3.1.6)$$

For n measurements of the log-normally distributed creep compliance J_0, the geometric mean of the measurements is derived by taking the arithmetic mean of the logarithmized data, which are normally distributed:

$$\overline{\ln J_0} = \frac{1}{n} \sum_{i=1}^{n} \ln J_{0,i}. \qquad (3.1.7)$$

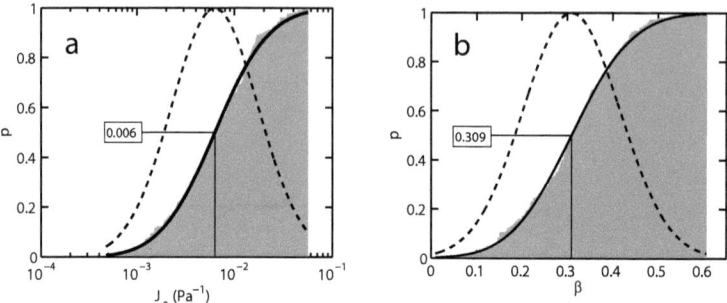

Figure 3.4: Cumulative histograms of the prefactor J_0 (a) and the exponent β (b). Data were pooled from 4 different cell lines (n = 185). Solid lines are fits to cumulative log-normal (a) and normal distributions (b), dotted lines denote the corresponding non-cumulative distributions. The median of the cumulative distributions corresponds to $p = 0.5$.

The geometric mean of the log-normally distributed data is then

$$\overline{J_0} = \exp\left(\overline{\ln J_0}\right), \tag{3.1.8}$$

and the standard error can be estimated by calculating the standard error in the logarithmic domain,

$$SE_{\overline{\ln J_0}} = \sqrt{\frac{1}{N(N-1)} \sum_{i=1}^{n} (\ln J_{0,i} - \overline{\ln J_0})^2}, \tag{3.1.9}$$

which gives in the linear domain

$$\overline{J_0} \pm SE_{\overline{J_0}} = \exp\left(\overline{\ln J_0} \pm SE_{\overline{\ln J_0}}\right) = \exp\left(\overline{\ln J_0}\right) \cdot \exp\left(\pm SE_{\overline{\ln J_0}}\right), \tag{3.1.10}$$

so the geometric mean has to be *multiplied* by the exponential standard error of the logarithmized data to obtain the standard error in the linear domain. All means and standard errors of creep compliance and shear modulus reported here were determined in this way.

3.1 Linear creep response

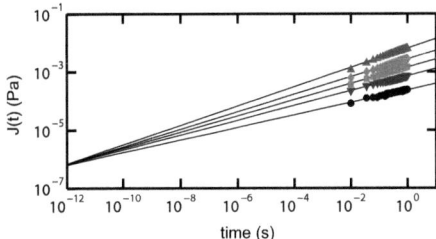

Figure 3.5: The creep curves for cells of different stiffness (n = 395) seem to exhibit a common intersection at small times. Data were binned by $1/J_0$. Solid lines are a fit to the master relation eq. (3.1.11)

3.1.4 Scaling the creep response

The creep curves for cells of different stiffness, when plotted with double-logarithmic axes, seem to exhibit a common intersection at small times (Figure 3.5). Consequently, they can be fit to the master equation

$$J(t) = j_0 \left(t/\tau_0 \right)^\beta, \qquad (3.1.11)$$

with a common intersection at j_0 and τ_0. This reduces the number of parameters to one, since the creep response is fully determined by the slope and the common intersection. The shear modulus as defined in the previous section is then not an independent parameter but can be derived from the power law exponent:

$$\ln(1/J_0) = \beta/j_0 \ln(\tau_0). \qquad (3.1.12)$$

This explains the different types of distributions of stiffness and exponent: a log-normal distribution of stiffness necessarily leads to a Gaussian distribution of the exponent, and vice versa.

The common intersection is most prominent in the ensemble average. In the individual data as shown in Figure 3.6, the dependence of J_0 on β is only apparent as a trend, which is probably due to variations of the measurement conditions such as the degree of bead embedding.

The master relation eq. (3.1.11) has also been observed in the frequency domain

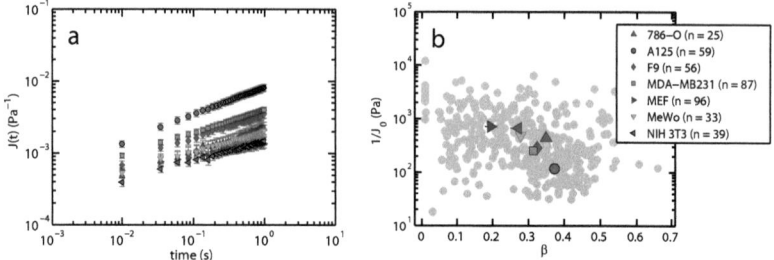

Figure 3.6: (a) Geometric mean of the creep response of mouse embryonal fibroblasts (MEF), NIH 3T3 mouse fibroblasts, F9 mouse embryonic carcinoma cells, MeWo human fibroblast-like cells, and MDA-MB231, 786-O and A125 human epithelial cancer cells. (b) Individual data for stiffness and power law exponent for all seven cell types. The relation between stiffness and power law exponent is visible as a trend in the raw data but becomes apparent in the ensemble average.

using magnetic twisting of beads [Fabry 03] or, very recently, by atomic force microscopy [Hiratsuka 09]. The linear rheology of adherent cells can therefore be characterized by a single master parameter: the power law exponent β. The implications of this observation and a more detailed interpretation of the exponent will be discussed in chapter 4.

3.1.5 Linearity and superposition

If the creep compliance $J(t)$ of a linear viscoelastic material is known, the Boltzmann superposition principle allows to predict the strain $\gamma(t)$ in response to an arbitrary time-dependent stress $\sigma(t)$:

$$\gamma(t) = J(t)\sigma(0) + \int_0^t J(t-t')\dot{\sigma}(t')dt'. \tag{3.1.13}$$

According to the Boltzmann formula, one expects for a force ramp $\sigma(t) = st$ and a power law creep compliance, $J(t) = J_0(t/t_0)^\beta$:

$$\gamma(t) = \int_0^t sJ_0 \left(\frac{t-t'}{t_0}\right)^\beta dt' = J_0 \frac{st_0}{\beta+1} \left(\frac{t}{t_0}\right)^{\beta+1}. \tag{3.1.14}$$

3.1 Linear creep response

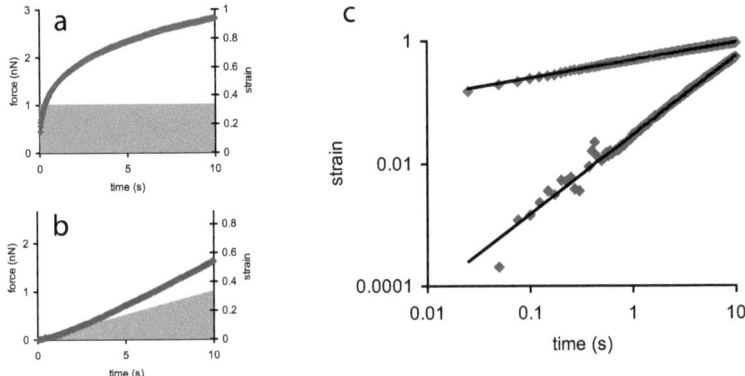

Figure 3.7: Average strain in response to a 1 nN constant force (a) and a linear increasing force ramp of the same maximum force (b). Same data in log-log representation are shown in (c), solid lines: power law fit with exponent 0.29 (red, creep) and 1.29 (green, ramp), as predicted from linear superposition.

The validity of linear superposition for small forces was assessed by measuring the bead displacement during application of a linear force ramp that increased up to 1 nN during 10 s. Since every creep measurement leads to irreversible changes of the cytoskeleton, step and ramp measurements were conducted independently on different cells of the same type, and the average displacements of both protocols were compared with the prediction from linear superposition.

As can be seen in Figure 3.7, theory and data were in good agreement for a force ramp up to 1 nN. Thus, for small forces and deformations and in a limited time domain, the cell's passive mechanical response is linear. This is in agreement with earlier studies of cell mechanics in the linear regime using other methods [Fabry 03, Feneberg 04, Lenormand 04].

3.1.6 Force dependence of parameters

The creep response to force steps of different magnitude always followed a power law, however both parameters were force-dependent. In general, the prefactor J_0 decreased with force (Figure 3.8a), indicating stress stiffening. The power law exponent showed

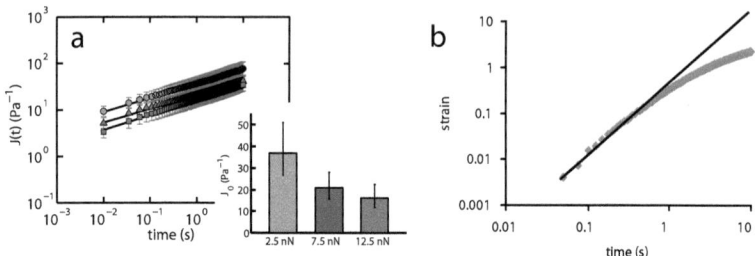

Figure 3.8: (a) Creep response to force steps of 2.5 nN (grey circles), 7.5 nN (green triangles) and 12.5 nN (blue squares) always followed a power law over time, but the decrease of J_0 with increasing force indicates stress stiffening (inset). (b) Strain in response to a 10 nN force ramp deviates from the power law behaviour predicted by linear superposition (solid line).

ambiguous force dependence.

A single force step is not a suitable protocol to determine the force dependence of the creep response, since in every measurement, only one force level is probed, and a large number of independent measurements on different cells is required to obtain statistically significant results. Additionally, effects of bead disruption, transient flow at high forces and geometric effects distort the results for high forces.

Upon application of a force ramp up to high force levels of up to 10 nN, the breakdown of superposition clearly became apparent. Significant deviations from the prediction of linear superposition were found (see Figure 3.8b), indicating that linearity does not hold and the creep response becomes nonlinear at high forces. The response to a ramp-like force in the nonlinear regime, however, cannot be interpreted without making further assumptions about the force dependence since time- and force-dependent properties cannot be disentangled.

Therefore, a novel approach to determine the time- and force-dependence of the creep response independently in a single measurement was developed and will be introduced in the following section.

3.2 Nonlinear differential creep

The force dependence of the creep response is investigated by applying a step force protocol. Stress stiffening (increase of $1/J_0$ with force) as well as fluidization (increase of β with force) are observed. Softer cells show a more pronounced stress stiffening, which is quantitatively explained by their smaller internal prestress. Stiffer and more elastic cells show a more pronounced force-induced fluidization, consistent with predictions from soft glassy rheology. Bead detachment is analyzed to obtain information about the adhesion strength of cells to extracellular matrix.

3.2.1 Differential step protocol

In order to quantify the force dependence of the creep response, a staircase-like sequence of increasing force steps was applied (Figure 3.9a), and the displacement after the n-th force step $d\sigma_n$ at time t_n was fit to a superposition of creep processes,

$$\gamma(t \geq t_n) = \gamma(t_n) + \sum_{i=0}^{n} [J_n(t-t_i) - J_n(t_u - t_l)] d\sigma_i, \quad (3.2.1)$$

with a response function $J_n(t)$ that depends not only on time, but also on the currently applied total force, $\sigma_n = \sum_{i=0}^{n} d\sigma_i$:

$$J_n(t) = J_0(\sigma_n) (t/t_0)^{\beta(\sigma_n)}. \quad (3.2.2)$$

Note that the creep response for all ongoing creep processes that started at some earlier time is according to the currently applied stress level at time t. Using the stress level at the retarded time $t - t'$ instead can result in negative values at higher forces if the ongoing old processes "get ahead" of the actual displacement. Moreover, oscillatory measurements of the shear modulus at different prestress in reconstituted cytoskeletal networks [Gardel 06a] indicate that the differential stiffness depends on the currently applied total stress, not on the history of force application or the timecourse of relaxation processes.

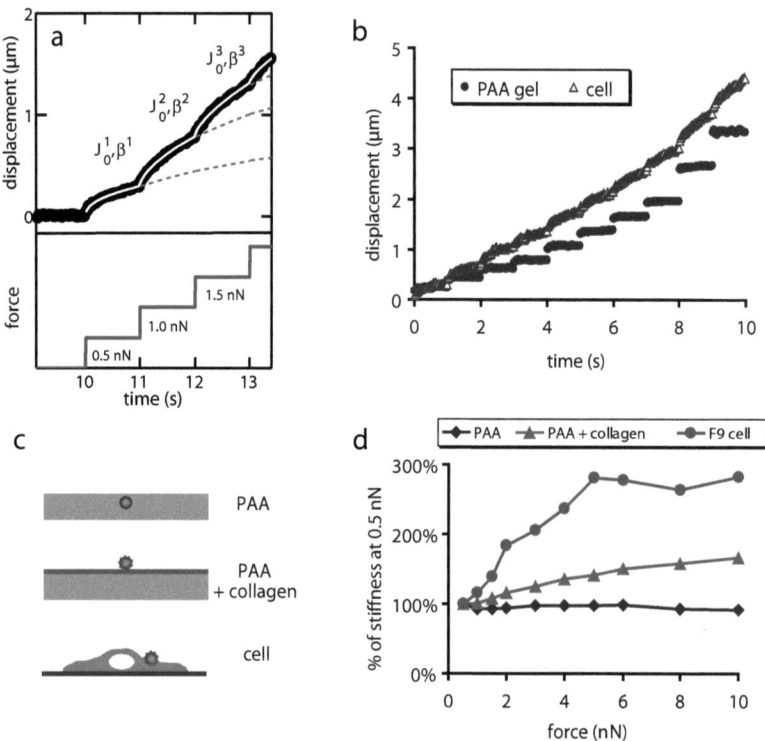

Figure 3.9: (a) A staircase-like sequence of increasing force steps was applied, and the displacement was fit to the corresponding superposition of creep processes. (b) Displacement of a bead embedded in PAA gel was elastic (closed circles), whereas a bead bound to a cell showed viscoelastic behavior (open triangles). (c) Beads were either embedded into the gel, or bound to the gel surface using fibronectin-coated beads and collagen-coated gels. (d) For PAA, stiffness was nearly independent of force for embedded beads, and slightly increased with force for surface-bound beads. Cells showed pronounced stress stiffening, in comparison.

3.2 Nonlinear differential creep

3.2.2 Stress stiffening and fluidization

The feasibility of the differential step protocol was assessed using polyacrylamide (PAA), which should behave as a linear elastic solid in the applied force range. Beads were either embedded into the gel, or bound to the gel surface using fibronectin-coated beads and collagen-coated gels (Figure 3.9c). A staircase-like sequence of ten force steps lasting one second each was applied, with forces ranging from 0.5 nN to 10 nN, and the resulting bead displacement was fit to the above equation.

As expected, the observed power law exponent was constantly close to zero, indicating purely elastic behaviour (Figure 3.9b). The resulting creep compliance J_0 was nearly independent of force for embedded beads, and slightly decreased with force for surface-bound beads (Figure 3.9d). This apparent stress stiffening is probably a geometric effect: beads have an additional rotational degree of freedom until the adhesion area is aligned in parallel to the direction in force, leading to a higher compliance at low forces. The same effect might contribute to the force dependence of the stiffness of cells, since the geometrical conditions are similar.

Subsequently, the same force protocol was applied to seven different adherent cell lines (mouse embryonal fibroblasts, NIH 3T3 fibroblasts, F9 mouse embryonic carcinoma cells, MeWo human fibroblast-like cells, and MDA-MB231, 786-O and A125 human epithelial cancer cells). Stress stiffening (increase of $1/J_0$ with force) as well as fluidization (increase of β with force) were observed in all cell types (Figure 3.10). Fibroblasts were on average stiffer but showed less stress stiffening compared to epithelial cells.

3.2.3 Prestress determines stress stiffening

To quantify the relationship between stiffness and stress stiffening, data from all experiments were pooled and binned by stiffness. As shown in Figure 3.11a, the stiffest and most elastic cells showed the smallest amount of stress stiffening.

The different degrees of stiffening can be explained by different levels of prestress in the cell: first, we assume that the total mechanical stress σ in the cytoskeleton is the sum of active internal prestress σ_p and passive external stress σ_e (Figure 3.12a). Furthermore, we assume that the linear relationship between differential stiffness $K'(\sigma)$ and cytoskeletal stress, which has been previously reported for smooth muscle cells [Wang 01] and reconstituted actin networks [Gardel 06a], probably also holds in the

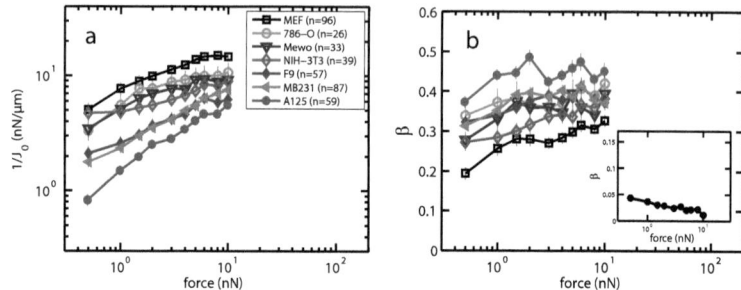

Figure 3.10: (a) Stiffness vs. force for seven different cell lines. Stress stiffening is more pronounced in soft cells (epithelial cells) compared to stiff cells (fibroblasts). (b) Power law exponent vs. force for same cell lines. Fibroblasts are initially more elastic and show more fluidization compared to epithelial cells. Inset: the small power law exponent of polyacrylamide, for comparison, indicates elastic behavior.

cell lines tested here. $K'(\sigma)$ can then be written as

$$K'(\sigma) = \frac{d\sigma}{d\gamma} = K'_0 + a(\sigma_p + \sigma_e), \tag{3.2.3}$$

where the unitless constant a characterizes the dependence of stiffness on stress, and K'_0 denotes the linear stiffness around the force-free state. Integration with the boundary condition $\sigma|_{\gamma=0} = 0$ yields an exponential stress-strain relationship, $\sigma = K'_0/a \left(e^{a\gamma} - 1\right)$ (Figure 3.12b). Such exponential relationships have been reported for whole cells, reconstituted cytoskeletal networks and many other biological tissues [Fernandez 08, Gardel 06a, Fung 93].

We fitted the force-stiffness curves to eq. (3.2.3) and found that the different amounts of stiffening are explained solely by σ_p (see Figure 3.11a). Accordingly, stiff cells have a more prestressed cytoskeleton, therefore the relative increase of mechanical tension and resulting stress stiffening due to external forcing is smaller than for soft cells.

The fit yields prestress values of up to 1500 Pa, which is in agreement with the maximum traction stress cells exert onto their substrate [Wang 01]. Interestingly, the value for K'_0 obtained from the fit, corresponding to the stiffness of the unstressed and prestress-free cell, is of order 5 Pa, similar to the linear shear modulus of crosslinked

3.2 Nonlinear differential creep

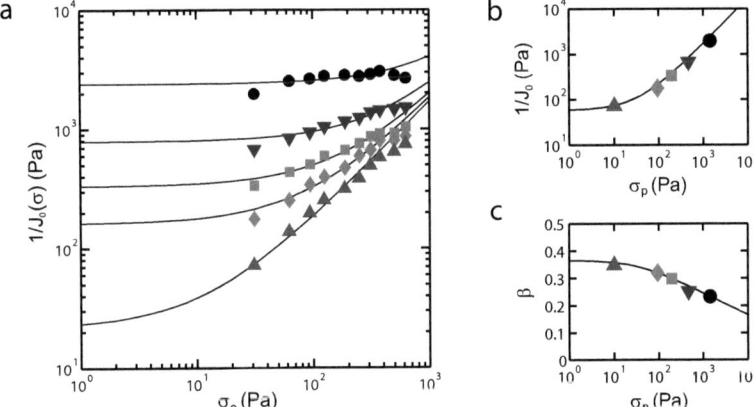

Figure 3.11: (a) Stiffness $1/J_0(\sigma)$ vs. applied stress σ_e, binned by the stiffness at the smallest force (n = 395). Stiff cells show less stress stiffening compared to soft cells. Black lines: fit to eq. (3.2.3) with common parameters $a = 1.68$ and $K'_0 = 5$ Pa, and prestress σ_p as free parameter. (b) Measured initial stiffness $1/J_0$ vs. σ_p for the same data. Solid line: prediction by eq. (3.2.3). (c) Measured exponent β vs. σ_p from Fig. 2a. Solid line: prediction by eq. (3.2.4). Standard errors are smaller than symbol size in all cases.

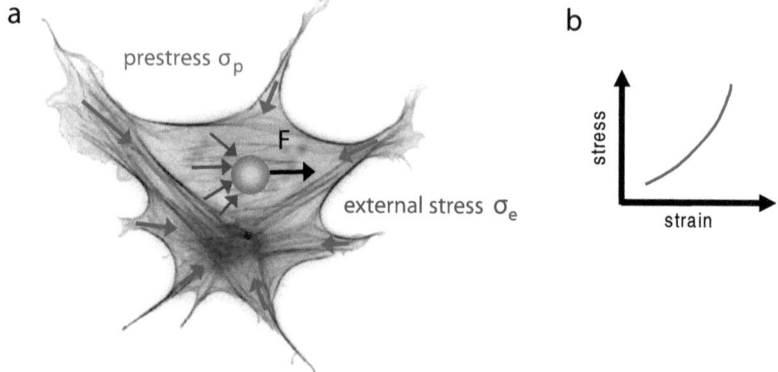

Figure 3.12: (a) The total mechanical stress σ in the cytoskeleton of an adherent cell is the sum of active myosin-generated internal prestress σ_p and passive external stress σ_e, e.g. induced through a bead. (b) The amount of actomyosin-generated prestress sets the position of the cell on the exponential stress-strain curve and thereby determines its differential stiffness.

actin networks [Gardel 06a].

These data suggest that the force dependence of the stiffness is coupled to the level of internal stress in the cell: the amount of actomyosin-generated prestress sets the position of the cell on the exponential stress-strain curve and thereby determines its differential stiffness.

3.2.4 Power law exponent and fluidization

Due to the common intersection of the linear creep curves and the corresponding master equation, linear stiffness and power law exponent are not independent variables, but are related via eq. (3.1.12). As has been suggested previously [Stamenović 04], for eq. (3.2.3) and eq. (3.1.12) to hold at the same time, the following relationship between prestress, stiffness and power law exponent must also hold:

$$\beta(\sigma_p) = \frac{\ln\left[j_0\left(K_0' + a\sigma_p\right)\right]}{\ln \tau_0}. \tag{3.2.4}$$

3.2 Nonlinear differential creep

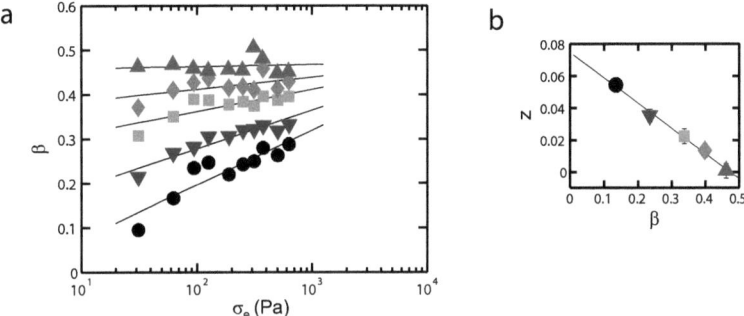

Figure 3.13: (a) Power law exponent β vs. applied stress σ_e. Data for all cell lines are binned by the exponent at the smallest force, β_0 (n = 395). Elastic cells (small β_0) show a higher amount of fluidization in response to force application compared to fluid cells (large β_0). Solid lines: fit to the empirical relation $\beta(\sigma_e) = \beta_0 + z \log(\sigma_e/\sigma_0)$. (b) The amount of fluidization, z, decreases linearly with the initial exponent β_0 (solid line).

This means that more contractile cells (higher prestress) are not only stiffer, they also display a smaller power law exponent and hence more solid-like properties compared to less contractile cells. The creep exponent obtained from the data in Figure 3.5, when plotted against the prestress obtained from the data in Figure 3.11a, closely follows eq. (3.2.4) (Figure 3.11c).

This relationship (eq. (3.2.4)), however, does not predict the behavior of β for higher external forces. The power law exponent β increased during force application, indicating force-induced fluidization and rupture events (Figure 3.13a). Cells with more solid-like behavior (small β) showed the most pronounced fluidization, whereas cells that were initially more fluid-like (large β) showed no further increase of β during creep (Figure 3.13b). Since β and prestress are related by eq. (3.2.4), cellular prestress plays again a central role as it determines the sensitivity of the cytoskeletal structure to external stress.

In smooth muscle cells, the more solid-like behavior under high prestress arises from reduced actomyosin cycling and energy dissipation due to a strongly bound latch state of actin and myosin [Fredberg 96]. This may apply to other contractile cells as well since they use the same mechanism of force generation. Under external force application,

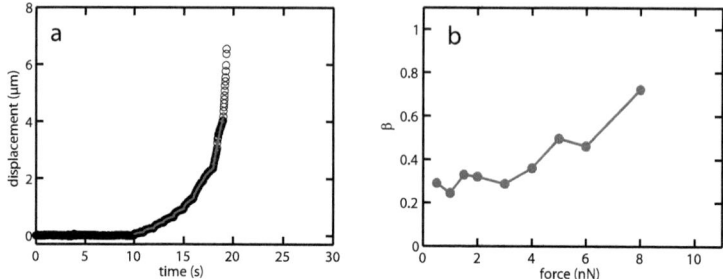

Figure 3.14: (a) Displacement of a representative bead that detached from the cell surface during force application. (b) Detachment is preceded by a flow period which manifests as an increase of the power law exponent.

these actin-myosin bonds store elastic energy, which increases their yielding probability. Force-induced rupture then leads to fluidization and increased dissipation, as seen by an increase of β. In cells with small prestress, on the contrary, the cytoskeleton flows in response to externally imposed stress, manifest in a large but force-independent β. A more elaborate explanation for such a mechanism is given in the subsequent chapter on theoretical modeling.

3.2.5 Quantification of adhesion strength

During force application, a fraction of beads detached from the cells. It is unknown where precisely in the mechanical chain between bead surface, ligand proteins, integrin receptors, focal adhesion complex and cytoskeleton the rupture takes place. Such detached beads were normally discarded prior to data analysis in order to avoid a bias in the rheological parameters. The force and time at which a bead detaches, however, reveal important information on the effective binding strength between bead and cell, and can be used to quantify differences in adhesion strength to extracellular matrix between different cell types.

Bead detachment is typically preceded by an increase of the power law exponent β as shown in Figure 3.14, corresponding to fluidization of the underlying structure. This coincides with the interpretation of the power law exponent as a measure for the turnover rate of force-bearing elastic elements as described by soft glassy rheology. Prior to detachment, the unbinding rate of the force-bearing elements increases and leads to

3.2 Nonlinear differential creep

a catastrophic avalanche of rupture events, which eventually results in yielding of the bead-cell link.

3.2.6 Force ramp in the nonlinear regime

From the results of the linear creep measurements described in section 3.1, it was not possible to predict the response to a force ramp (Fig. 3.8b). The reason is that the parameters J_0 and β of the power law creep response depend on force and hence, superposition does not hold. Now that the force dependence has been determined from the nonlinear step-creep measurements, it should be possible to fit the nonlinear bead displacement during a force ramp.

To fit the bead displacement due to a linear force ramp, the formula eq. (3.2.1) on page 51 is used, with several simplifications due to the force protocol that are explained in the following. The change of strain between times t_n and $t_n + dt$ is given by the second term on the right side of eq. (3.2.1),

$$d\gamma(t_n) = \sum_{i=0}^{n} [J_n(t_n + dt - t_i) - J_n(t_n - t_i)] d\sigma_i = \sum_{i=0}^{n} dJ_n(t_n - t_i) d\sigma_i, \quad (3.2.5)$$

where the sum runs over all ongoing creep processes due to past force steps σ_i. During experiments, the bead displacement is measured in discrete time intervals dt, so that $t_n = i dt$. Furthermore, the force increase between two measured timepoints in the ramp case is constant and time-independent, $d\sigma_i = const. = d\sigma$. The total strain at time t_N, which is the sum over all strain changes, can then be written as

$$\gamma(t_N) = \sum_{n=0}^{N} d\gamma(t_n) = d\sigma \sum_{n=0}^{N} \sum_{i=0}^{n} dJ_n[(n-i)dt] = d\sigma \sum_{n=0}^{N} \sum_{m=n}^{0} dJ_n(mdt), \quad (3.2.6)$$

where the index of the second sum has been replaced by $m = n - i$. The bead displacement in response to a force ramp can therefore be fit by the equation

$$\gamma(t_N) = d\sigma \sum_{n=0}^{N} J_n(t_n) \quad (3.2.7)$$

with the sum over all creep responses J_n at timepoints t_n and total force level σ_n.

The bead displacement of 31 beads bound to NIH 3T3 cells in response to a force that increased linearly from 0 to 10 nN in 10 seconds was measured and fit to eq. (3.2.7), with a force-dependent creep response $J(t,\sigma)$ according to eq. (3.2.3) and Fig. 3.13:

$$J(t,\sigma) = \left[K_0' + a(\sigma_p + \sigma_e)\right]^{-1} \left(\frac{t}{t_0}\right)^{\beta_0 + z \log(\sigma_e/\sigma_0)} \tag{3.2.8}$$

The resulting fit parameters were then compared to the values of $1/J_0$ and β obtained previously by step-creep measurements (Fig. 3.15). The increase of stiffness with force in the ramp experiments (Fig. 3.15b) is within errors identical to that obtained in creep experiments. This shows that eq. (3.2.8) correctly describes the force dependence of the elastic properties of cells.

The power law exponent β that results from the ramp fit increases with force as in the creep experiments, but with higher initial values and smaller slope than observed during creep (Fig. 3.15c). Since β indicates the time dependence of the cell rheological properties, the observed difference might reflect the different time course of force in both protocols. This could be clarified in future experiments by applying force ramps at different rates and observing how β changes with the rate of force application.

3.2 Nonlinear differential creep

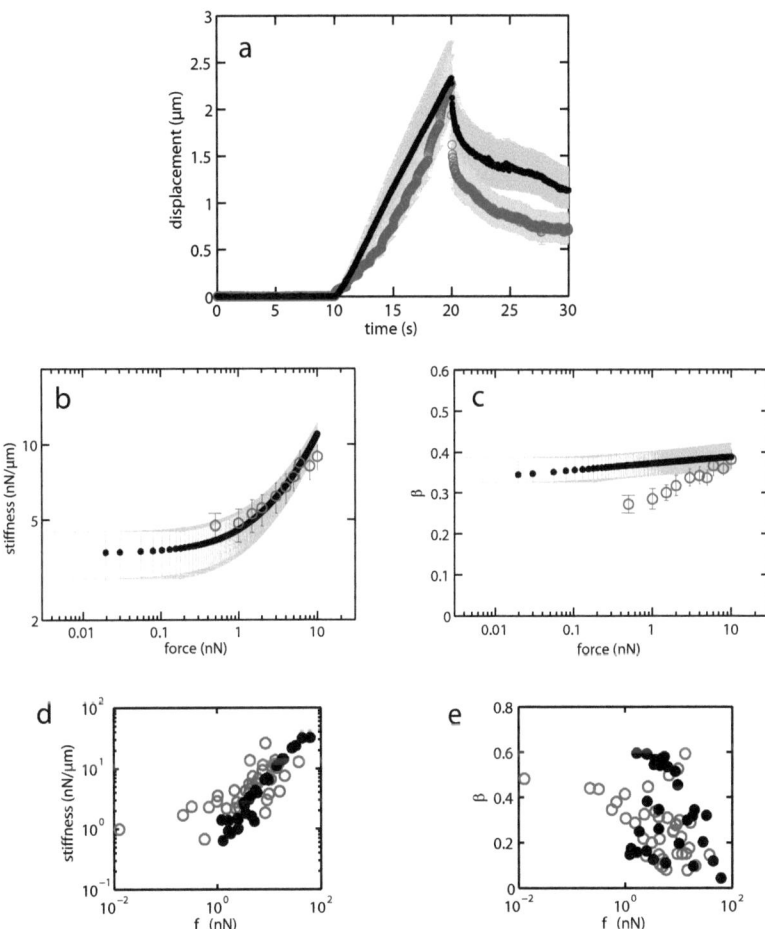

Figure 3.15: (a) Mean displacement of beads bound to NIH-3T3 fibroblast cells in response to a 10-second force ramp with a maximum force of 10 nN (black dots, n=31) and in response to a step force as described in section 3.2.1 (red circles, n=39). (b) Stiffness vs. force and (c) power law exponent vs. force for the force ramp (black dots) compared to the step force (red circles). (d) Stiffness vs. prestress and (e) power law exponent vs. prestress derived from the fit of eq. (3.2.8) to the ramp (black dots) and step (red circles) data.

3.3 Creep recovery and plasticity

In this section, the amount of permanent plastic deformation in the cytoskeleton due to creep is investigated. To get quantitative results, the recovery of the bead displacement after switching off the force was compared to the prediction of linear superposition. Recovery was at the same time incomplete and more fluid-like, indicating structural changes and a speed-up of dissipative processes in the cytoskeleton. This reflects persistent changes of the mechanical properties of the underlying cytoskeleton, which was assessed by applying repeated force pulses. A steady state was typically reached after about five cycles of creep and recovery, reminiscent of preconditioning observed in smooth muscle and biological tissue.

3.3.1 Quantification of creep recoil

In the single-step creep experiments described in the first part of the experimental results in section 3.1, a constant force was applied to a bead for ten seconds, and then switched off. After switching off the force, the bead displacement partially recovered, thereby releasing the stored elastic energy. For any viscoelastic material, the creep deformation is expected to recover only partially since a fraction of the imposed mechanical energy has already been dissipated during creep. To quantify the creep recovery, or "degree of recoverable strain" as it is called in engineering literature [Shenoy 02], the bead displacement was recorded for another ten seconds after switching off the force, and compared to the prediction of linear viscoelasticity.

The timecourse of creep recovery for a linear viscoelastic material is predicted by the constitutive equation of the material. According to linear superposition, force cessation after ten seconds is equivalent to continued force application and onset of an additional force of equal magnitude, but opposite direction (Fig 3.16a). The expected timecourse of creep recovery is then simply the superposition of the continuing creep process and the new creep process in response to the opposing force (Fig. 3.16b).

The actual measured displacement during recovery is fit by the corresponding superposition of the old and new power law creep response. In order to quantify the deviation of the cytoskeletal creep recovery from the prediction of linear superposition, the two parameters of the power law are allowed to differ from those of the preceding creep. The quotients of the predicted and actual prefactors and power-law exponents are then used to quantify the amount of permanent structural changes induced by the creep de-

3.3 Creep recovery and plasticity

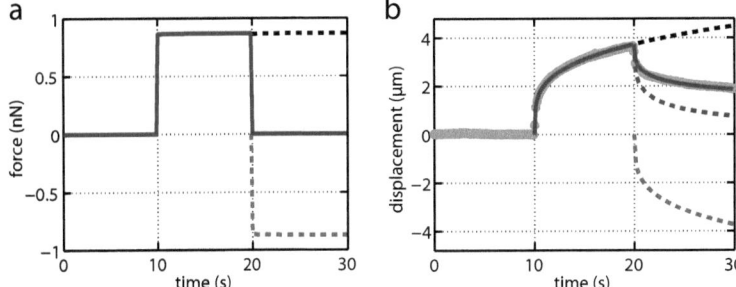

Figure 3.16: a) The force protocol applied during a creep experiment is shown as a solid blue line. A total force level of zero during recovery is equivalent to a superposition of two forces of equal magnitude, but opposite direction (dashed black and red lines). b) Accordingly, the expected displacement after force cessation (dashed blue line) is equivalent to the superposition of the ongoing creep (dashed black line) and a new creep process in the opposite direction (dashed red line). The actual displacement deviates from this prediction, but can nevertheless be fit by a superposition of two power laws (solid blue line) with new parameters. Grey circles are data points for one representative measurement.

formation.

3.3.2 Incomplete recovery

The amount of recovered strain is expressed by the quotient $R = J_r/J_0$ of the prefactors J_r for recovery and J_0 of the foregoing creep process. It denotes the ratio of measured to predicted displacement or, if R is multiplied by 100, the percentage of recovery one second after switching off the force. A value of R smaller than one means that the bead displacement recovers less than predicted by linear superposition, or the amount of permanent remaining deformation one second after force cessation is higher than expected.

The recovery R for 147 individual measurements is shown in Fig. 3.17a. In most cases, R is smaller than one, corresponding to incomplete recovery. R shows a negative dependence on the preceding creep compliance, J_0: the more the bead has been displaced from its original position after ten seconds of creep, the smaller was the percentage of recovery. Apparently, the amount of strain is directly linked to the degree of permanent structural changes and plastic deformation induced in the cytoskeletal structure during creep.

3.3.3 Increase of power law exponent

The superposition principle does not only predict the amount, but also the time dependence of recovery, expressed by the power law exponent β. Again, we define a quotient $S = \beta_r/\beta$ of the power law exponents β_r for recovery and β for creep. A ratio S larger than one means that the creep recovery shows a more fluid-like behavior compared to the preceding creep. A more fluid-like response corresponds to a higher rate of energy dissipation and faster relaxation processes in the underlying structure. We therefore denote the ratio S of the power law exponents as the amount of speed-up after creep.

The measured speed-up ratio S was larger than one in most, but not all cells. When S is plotted against the initial power law exponent β as shown in Fig. 3.17b, it becomes clear that the speed-up is most pronounced for the initially most solid-like cells, whereas initially more fluid-like cells tend to have speed-up ratios S smaller than one, corresponding to a decrease of the power law exponent with respect to the preceding creep.

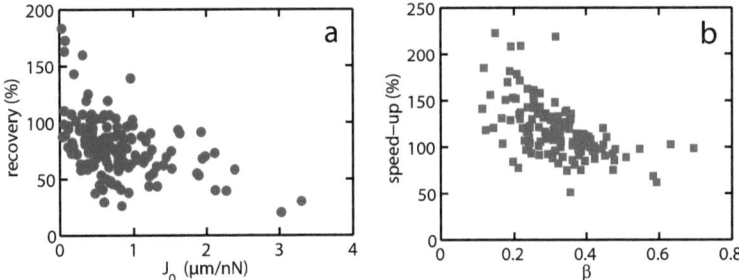

Figure 3.17: a) Recovery ratio $R = J_r/J_0$ versus creep compliance J_0 for MDA-MB 231 breast carcinoma cells (n−147), expressed as percentage. The recovery is typically around 100% for the stiffest cells and decreases with increasing compliance. b) Percentage of speed-up, determined from the ratio $S = \beta_r/\beta$ of the power law exponents, for the same cells. Initially more elastic cells show an increased power law exponent during recovery, whereas initially more fluid cells tend towards a decrease of the power law exponent during recovery.

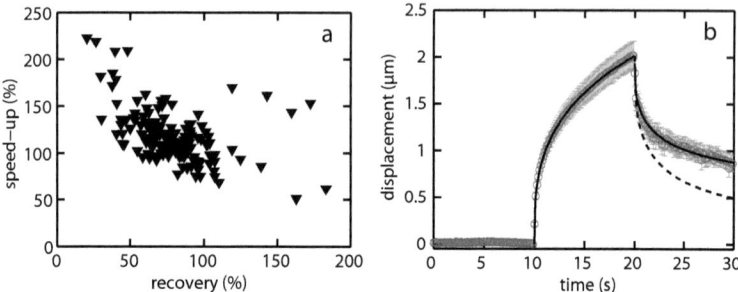

Figure 3.18: a) Speed-up $S = \beta_r/\beta$ versus recovery $R = J_r/J_0$ for MDA-MB231 breast carcinoma cells (n=147), expressed as percentage. Cells that recover less show at the same time a more pronounced increase of the power law exponent. b) Geometric mean of the displacement of all 147 cells. The error bars denote the standard error of the mean. Black solid lines are the power law fits to creep and recovery, whereas the dashed black line shows the expected recovery predicted by linear superposition.

3.3.4 Prestress determines recovery and speed-up

It has already been shown in section 3.1.4 that the two parameters of the power law creep response in cells are not independent. The prefactor J_0 and the exponent β are related via the master equation eq. (3.1.11). Since the recovery and speed-up parameters R and S show a dependence on J_0 and β, respectively, it can be concluded that they are themselves not independent. When S is plotted against R, the relationship between the two clearly becomes apparent (Fig. 3.18a). Cells that recover less show at the same time a more pronounced increase of the power law exponent.

On average, recovery was incomplete ($\langle R \rangle = 0.76$) and more fluid-like ($\langle S \rangle = 1.14$), as summarized in Fig. 3.18b. The distributions of R an S have been determined from 147 independent measurements on the same cell type. Both parameters follow a log-normal distribution (Fig. 3.19), in contrast to the normally distributed power law exponent that underlies the speed-up ratio, S.

In section 3.2.4 of the experimental results, it has been elucidated that the linear and nonlinear creep response is essentially determined by a single master parameter: the cytoskeletal prestress, σ_p. From the results of this section, it can be concluded that if the two power law parameters J_0 and β of the linear creep response depend on pre-

3.3 Creep recovery and plasticity

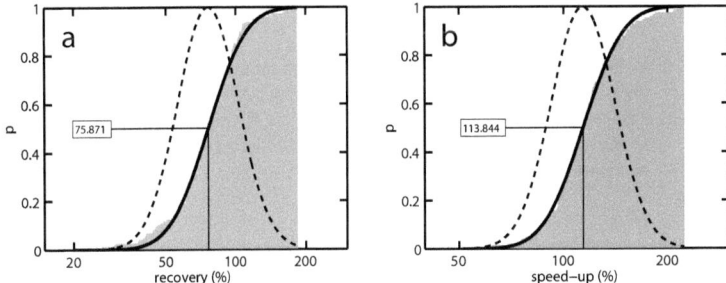

Figure 3.19: Cumulative histograms of the percentage of recovery (a) and speed-up (b) after ten seconds of creep for MDA-MB231 cells (n=147). The solid lines denote the fit to a cumulative log-normal distribution, whereas the dashed lines show the corresponding non-cumulative distributions. Both parameters follow log-normal distributions, despite the normally distributed power law exponents underlying the speed-up ratio. The numerical values in the boxes denote the value at $p = 0.5$, corresponding to the median of the distribution.

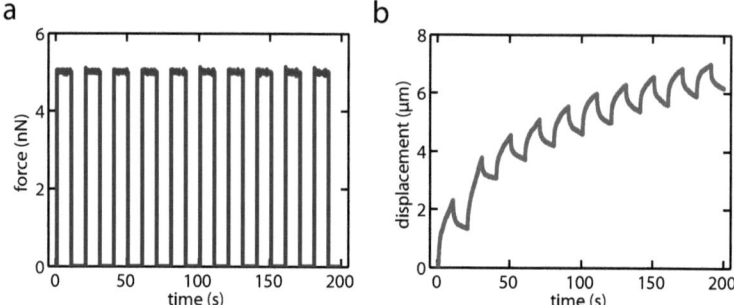

Figure 3.20: a) Repeated application of a step force to characterize the development of the mechanical response during alternating cycles of creep and recovery. b) Displacement of a bead during application of the force protocol shown in a). Every creep and recovery period is fitted by a respective power law with two parameters.

stress, then so does the recovery of creep and the amount of permanent plastic deformation. This contributes to the notion that elastic, viscous and plastic properties of the cytoskeleton are different aspects of the same underlying mechanism of energy storage and dissipation, which motivates the theoretical model described in chapter 4.

3.3.5 Repeated force steps and preconditioning

If the incomplete creep recovery is caused by permanent structural changes in the cytoskeleton, this should also affect the mechanical properties after creep. Therefore, the results of a second creep measurement that is started 20 seconds after the first one should reflect these changes. To verify this expectation, repeated creep measurements were performed by applying alternating force on-off pulses of identical duration and magnitude (Fig. 3.20).

The resulting prefactors and power law exponents of the alternating creep and recovery periods are shown in Fig. 3.21. It becomes apparent that a kind of steady state is reached after about five cycles. After that, the creep parameters do not change any more between subsequent cycles of force application and recovery. This observation is reminiscent of preconditioning of tissue or smooth muscle: prior to rheological measurements on such materials, repeated cycles of stretch and release are applied until a steady

3.3 Creep recovery and plasticity

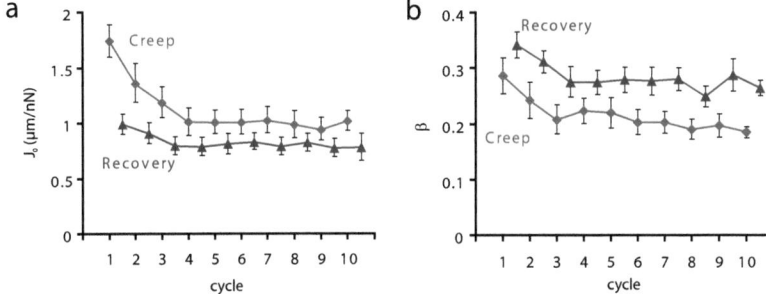

Figure 3.21: a) Creep compliance J_0 versus cycle for alternating creep and recovery. Cells become on average stiffer and develop a steady state after five cycles. b) Power law exponent β versus cycle. The recovery dynamics is consistently more fluid than the respective creep, and both develop towards a more solid-like steady state.

state is reached in order to obtain reproducable experimental conditions, as described by Fung [Fung 93]. According to this description, preconditioning is usually associated with changes of the internal structure and rupture of bonds during repeated cycling until a steady state is reached where no further internal changes occur.

3.4 Biological applications

The high-force magnetic tweezers method and the nonlinear step-creep protocol developed in the course of this work have been applied to answer a number of cell biological and biomedical questions. The role of the focal adhesion protein vinculin as a mechanoregulator has been studied. Results of experiments on cells lacking the actin crosslinker Filamin-A (FLNa) revealed that FLNa is responsible for the active, but not for the passive stress stiffening observed in cells. Furthermore, the role of cell mechanics for cancer cell invasion into connective tissue has been studied.

3.4.1 Vinculin as mechanoregulator

An important application of cell rheology is to determine the influence of different structural proteins on the mechanical properties of cells. As an example, the influence of the focal adhesion protein vinculin was studied by comparing the mechanical properties of cells lacking vinculin, or expressing only parts of the protein, to wild-type controls [Mierke 08a]. Creep measurements in the linear regime were carried out on F9 wild-type cells and $\gamma 229$ vinculin knock-out and rescue cells (Fig. 3.22a). Cells lacking vinculin showed reduced stiffness and an increase of the power law exponent, as shown in Fig. 3.22b.

The adhesion strength of cells to extracellular matrix was determined by quantifying the bead rupture during step-creep experiments (Fig. 3.22c). Bead binding strength was reduced in the cells lacking vinculin or expressing only the head fragment of the protein. These findings point towards a prominent role of the vinculin tail for mechanical stability and transduction of mechanical signals. Further studies are now being carried out to investigate the importance of the lipid binding motif residing on the vinculin tail, and the role of phosporylation.

3.4.2 FLNa determines active but not passive stiffening

In-vitro experiments on reconstituted actin networks revealed that the flexible actin crosslinker Filamin-A (FLNa) plays a crucial role for the stress stiffening of crosslinked actin networks that serve as a model system for the cytoskeleton [Gardel 06a]. The unique property of FLNa to unfold sequentially under force seems to be a key factor, since actin networks crosslinked by FLNa mutants lacking the unfolding domain did

3.4 Biological applications

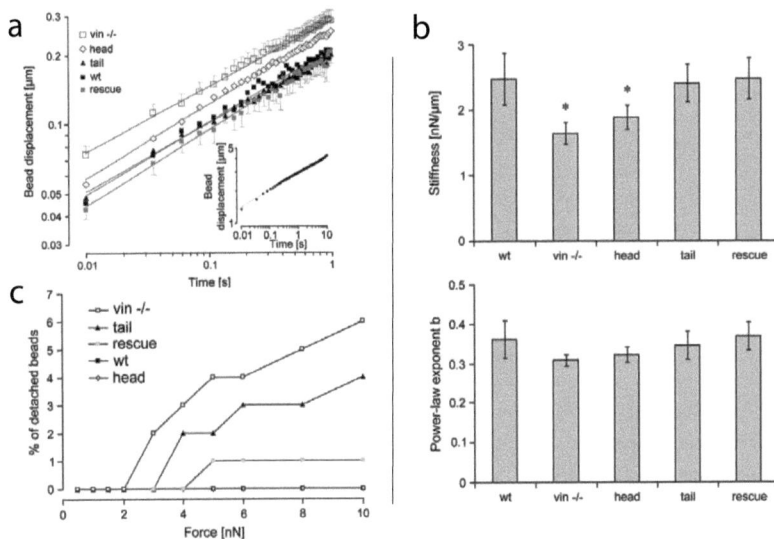

Figure 3.22: a) Creep response (geometric mean of bead displacement over time + geometric SE) to a 0.5 nN force step. The creep response for all cells followed a power-law relationship over two time decades and differed between F9 wild-type and the vinculin mutant cell lines. Between 60 and 86 cells from each cell line were measured. (Inset) Creep response of a representative F9 wild-type cell measured over three time decades (0.01-10 s). The solid lines show the power-law fit to the data. b) Stiffness (top row) and power-law exponent (bottom row) in F9 wild-type and vinculin mutant cell lines obtained from the power law fit to the creep response to a 0.5 nN force step ($p \leq 0.05$.). c) The percentage of detached beads versus force for the F9 wildtype and the vinculin mutant cell lines. Between 60 and 86 cells from each cell line were measured. Reproduced from [Mierke 08a].

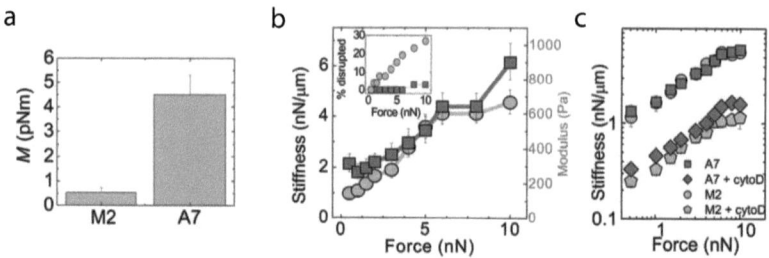

Figure 3.23: a) Stiffening of A7 and M2 cells subjected to large nanonewton-scale external forces, applied by magnetic tweezers through a fibronectin-coated magnetic bead. (A) Cell stiffness and estimated modulus (geometric mean ± SE) as a function of force for A7 (squares, n = 30) and M2 (circles, n = 19) cells. (Inset) Percentage of beads disrupted from cells as a function of force. (B) Cell stiffness as a function of force for A7 (n = 9) and M2 (n = 7) cells treated with 5 mM cytochalasin-D for 15 min. From [Kasza 09].

not exhibit comparable amounts of stress stiffening [Gardel 06b]. Numerical simulations of networks crosslinked by unfolding links support this hypothesis [Hoffman 07, DiDonna 06]. This has inspired the search for comparable effects of FLNa in cells.

In an extensive study [Kasza 09], the active and passive mechanical properties of FLNa-deficient melanoma cells (M2) were compared to a variant of that cell type called A7 where FLNa had been reintrocuded. Whereas the active prestress the cells generate and their ability to adapt mechanically to substrates of different stiffness differed significantly (Fig. 3.23a), the differences in the linear creep response and dynamical shear modulus were much less pronounced. The force dependence of the creep modulus was identical in both cell types, as shown in Fig. 3.23b. The inset of that figure shows, however, that the bead binding strength was much higher in the cells where FLNa was present.

Interestingly, the disruption of the actin cytoskeleton by application of 5 μM cytochalasin-D for 15 minutes resulted in a significant softening of both cell types, whereas the force dependence remained the same. This result seems to contradict the finding that the degree of stress stiffening depends on the differential stiffness and is determined by the amount of prestress in the cell (section 3.2.4). It indicates that not only the actin network is responsible for stress stiffening, but also other filamentous components of the cytoskeleton such as the intermediate filament network.

3.4 Biological applications

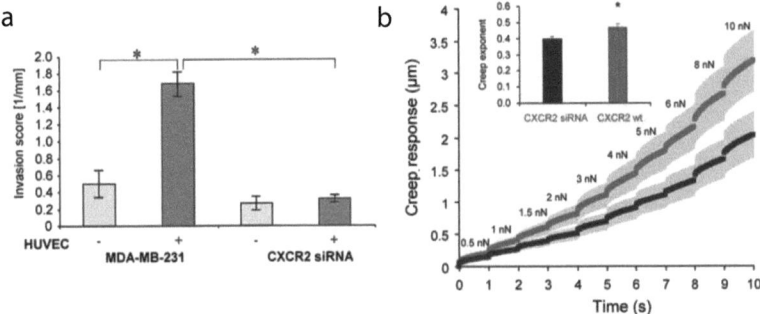

Figure 3.24: a) In cells where the chemokine receptor CXCR2 has been silenced using si-RNA, the invasion into the extracellular matrix is not enhanced by an additional layer of endothelial cells. b) The reduced invasivity due to knockdown of CXCR2 is accompanied by a decrease of the power law exponent and a general reduction in creep compliance. From [Mierke 08b].

Furthermore, the fact that the linear and nonlinear rheological properties in both cell types are almost identical despite the differences in prestress is a sign that the two cell types also differ substantially in their internal structure, hindering a direct comparison of their mechanical properties. Among individual measurements of each cell type, the relationships between linear stiffness and stress stiffening as well as between linear power law exponent and fluidization as described in section 3.2 remained valid.

The role of intermediate filaments for cell rheology is currently investigated in an ongoing study using various biochemical approaches to influence this component of the cytoskeleton. The M2 and A7 cells are a good model system for these studies since they are known to express large amounts of the intermediate filament vimentin, which might contribute to a higher degree to their mechanical properties as previously assumed.

3.4.3 Cancer cell metastasis and cell rheology

In the process of cancer cell metastasis, individual cells are believed to separate from the primary tumor and to invade the surrounding extracellular matrix, eventually reaching the blood circuit of the organism which transports them to other organs, where they complete their fateful cycle by invading again into the tissue and establishing a secondary tumor. The metastatic potential of a tumor is the key parameter that indicates the letality of cancer.

Migration of a tumor cell through the extracellular matrix requires the cell to generate forces in order to deform the matrix, and to adapt its own shape in order to squeeze through the dense mesh of collagen fibers. The mechanical properties of the cell play an important role in this context. Rheological studies on cancer cells contribute to the understanding of the process of metastasis and could be a possible diagnostic tool to quantify the metastatic potential of these cells.

The magnetic tweezer device built in the course of the work is used extensively in ongoing studies of cell microrheology in conjunction with tumor cell transmigration. As an example, the results of a study linking the mechanical properties of tumor cells to the expression of the CXCR2 in these cells is shown in figure 3.24. CXCR2 is a receptor for the chemokines IL-8 and Gro-β, which are secreted by endothelial cells and are believed to stimulate the transmigration of tumor cells through the endothelial layer of blood vessels. The expression of CXCR2 correlates strongly with the creep compliance as well as with the invasion through the endothelial layer in an in-vitro transmigration assay [Mierke 08b]. In cells expressing low levels of CXCR2, invasion into the extracellular matrix is not enhanced by the presence of an endothelial layer. At the same time, these cells show a decreased power law exponent and a general reduction in creep compliance compared to the untreated wild-type cells.

A more general hypothesis that is still under investigation is that the power law exponent, which can be interpreted as a measure for cytoskeletal dynamics and the effective matrix temperature of the cytoskeleton, is higher in invasive cells that need to be more active and dynamic in the course of migration compared to non-invasive cells.

4 Theoretical Model

4.1 Introduction

In this section, the groundwork is laid for a comprehensive theoretical model of cell mechanics that reconciles the phenomenological approach of Soft Glassy Rheology (SGR) with mechanistic descriptions of filament elasticity, bond stability, and force generation. First, SGR is reviewed in the context of cell mechanics, followed by introductions into the Wormlike Chain (WLC) theory of semiflexible filaments, the Bell model of force-dependent unbinding, and the Sliding Filament (SF) theory of acto-myosin force generation. Building on these descriptions, the basic idea of the proposed model is motivated and introduced. Details of the model and its numerical implementation will then be described in the subsequent section.

4.1.1 Overview

In the experimental part of this work, the nonlinear creep response of living cells was characterized. The essential results are that power law creep holds at high forces, and that stress stiffening and fluidization are universal and depend on the amount of prestress in the cell. These findings contribute to the observation that, despite the enormous complexity of the underlying structures, cellular microrheology seems to obey few, clearly defined empirical laws [Trepat 08]. Many of these phenomena are as well observed at higher hierarchical levels such as whole cells [Desprat 05, Fernandez 06, Fernandez 08], cell compounds [Fernandez 07], and even tissue [Fung 93, Fredberg 89]. This suggests that the mechanical behavior of these systems arises from common principles that do not depend on the different structural details of the constituents.

Several attempts have been made to define such principles and to derive a comprehensive model of cell and tissue mechanics, yet each of these approaches explains only a limited subset of observations. A mechanistic picture that captures the full phenomenology would help us to learn how structural and molecular mechanisms contribute to cell rheology, and vice versa, how molecular mechanisms can be understood from cell rheology measurements.

In the subsequent sections, the phenomenological way of describing cells as soft glassy materials will be contrasted with mechanistic descriptions of filament elasticity, bond dynamics and force generation. The introduction concludes with a proposal how these contrasting approaches can be combined into a quantitative model that explains all relevant observations with the minimum amount of complexity.

4.1.2 Soft Glassy Rheology (SGR)

Already when the first rheological experiments on biological specimens were conducted in the 19th century, it was reported that their creep response and stress relaxation do not decay with one or few characteristic time constants, but show a long-lasting logarithmic or power-law time dependence [Weber 35]. The results of creep measurements on living cells described in the experimental results of this work and in earlier studies using different techniques [Fabry 01, Hoffman 06] substantiate these findings: The linear creep response $J(t)$ or, correspondingly, dynamic shear modulus $G(\omega)$ under small deformations follows a weak power law over several orders of magnitude of time or frequency.

This timescale-invariance or lack of characteristic relaxation times has inspired the notion of cells as soft glassy matter close to a glass transition, similar to other soft disordered materials with power law moduli, such as foams, slurries or colloids [Fabry 01]. Such materials have in common that they are disordered and metastable, with the result that thermal motion is not sufficient to achieve structural relaxation.

A class of theories by which such systems are described in soft matter physics are called *trap models*: elements of the system "see" a landscape of energy traps of varying depth defined by their neighbouring elements [Bouchaud 92]. Internal dynamics and rearrangements of the system correspond to thermally activated hopping of elements between these energy traps. As the temperature is lowered, thermal energy is no more sufficient for elements to hop out of their traps, leading to kinetic arrest and a glass transition.

The cooperative effect of rearrangements on the hopping between traps is expressed by an effective noise temperature, x, that replaces $k_B T$ in the reaction rates. Continuous deformation and yielding leads to a sawtooth-like local element strain, as predicted e.g. for ideal foams. If the elements behave as ideal Hookean springs and store elastic energy under deformation, this energy gets dissipated as the element hops into another trap (Fig. 4.1a). The maximum elastic energy stored up to the yield length l is then $E = \frac{1}{2}kl^2$, with a spring constant k.

The master equation for the probability $P(E, l, t)$ of finding an element with yield energy E and local strain l at time t is

$$\frac{d}{dt}P(E,l,t) = -\dot\gamma \frac{d}{dl}P - \Gamma_0 e^{-[E-\frac{1}{2}kl^2]/x} P + \Gamma(t)\rho(E)\delta(l) \qquad (4.1.1)$$

with $\dot\gamma$ the global shear rate, Γ_0 an attempt frequency for hops, and $\Gamma(t)$ the total yielding

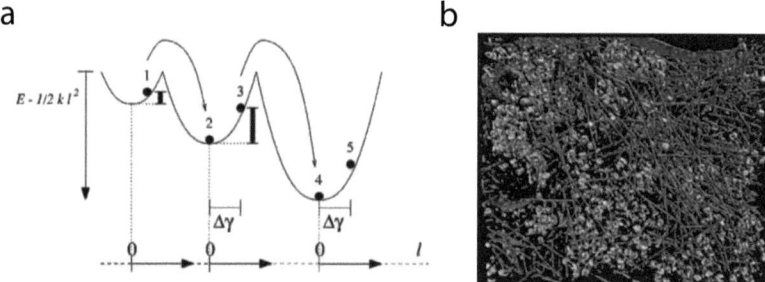

Figure 4.1: a) Potential well picture of the dynamics of the Soft Glassy Rheology (SGR) model, from [Sollich 98]. The relative horizontal displacement of the potential wells is arbitrary; each has its own independent zero for the scale of the local strain l. The solid vertical bars indicate the energy dissipated in the yield events from 1 to 2 and 3 to 4, respectively. b) Vizualization of actin network (red), membranes (blue) and ribosomes (green) in a 815 nm by 870 nm by 97 nm cell volume using cryo-electron microscopy and surface rendering, from [Medalia 02]. The crowded heterogenous structure of the cell interior supports the analogy to soft glassy materials.

4.1 Introduction

rate which is given by

$$\Gamma(t) = \Gamma_0 \left\langle e^{-[E-\frac{1}{2}kl^2]/x} \right\rangle_P = \Gamma_0 \int dEdl P(E,l,t) e^{-[E-\frac{1}{2}kl^2]/x}. \quad (4.1.2)$$

From this master equation, rheological constitutive equations can be derived. For example, the stress over time in a shear rate experiment is

$$\sigma(t) = k \int dEdl P(E,l,t) l. \quad (4.1.3)$$

If the trap depths follow an exponential distribution, $\rho(E) = \exp(-E)$, corresponding to a Gaussian distribution of yield strains, the model exhibits different power law regimes of the viscoelastic moduli for different values of the effective temperature x. For $1 < x < 2$, the storage and loss moduli $G'(\omega)$ and $G''(\omega)$ scale with frequency as ω^{x-1}, leading to the weak power law rheology of soft glassy materials observed e.g. in small-amplitude dynamic oscillatory measurements. For $x \to 1$, the system becomes more and more solid-like and approaches a glass transition.

Although it seems far-fetched at first to identify living cells with glass-like soft materials, there are a number of striking similarities that support this analogy. First of all, the heterogenous disordered structure of the cytoskeleton and the crowded environment of the cell is not much different from collodial or other crowded heterogenous materials, as shown in Fig. 4.1b. Moreover, cells have been shown to display a number of nonlinear "glassy" characteristics such as aging and rejuvenation, shear fluidization, and yielding [Bursac 05, Trepat 07]. The increasingly more fluid-like creep response with increasing force reported in the previous chapter is also consistent with SGR.

There are, however, a number of observations that SGR cannot explain: soft glassy materials generally show shear thinning and yielding under large stresses. The stress stiffening behavior that is characteristic of most biological materials cannot be described by SGR. Moreover, there is no mechanism of internal force generation by active elements: after yielding, elements always restart from a zero-strain force-free state.

Despite these limitations, SGR is a compelling analogy and has inspired many innovative studies in cell mechanics. In the field of airway physiology, the SGR hypothesis even lead to useful insights about the role of smooth muscle reorganization in asthma [Fabry 03]. However, it remains an analogy on a phenomenological level with limited predictive power. To date, no mechanistic explanation has been given which constituents of the cytoskeleton correspond to the elements of the SGR model, or what the noise temperature x means in terms of physical quantitites. To put cytoskeletal rheology onto a more physical basis, three models that explain different aspects of cell mechanics in a quantitative mechanistic way are introduced in the following sections.

4.1.3 The Wormlike Chain (WLC)

The reductionist method of physics, in contrast to the phenomenological approach used in the previous section, explains the properties of a system from the properties of its constituents. In the case of cell mechanics, this bottom-up strategy means to start with quantitative in-vitro studies of the constituents of the cytoskeleton, and then to link together the functional modules into higher-order structures with emerging biological complexity [Bausch 06].

The cytoskeleton of eukaryotic cells is a crosslinked network of semiflexible filaments that spans the entire cell. We therefore start with the minimum model for semiflexible filaments: the Wormlike Chain (WLC).

Semiflexible polymers are characterized by a persistence length ℓ_p that is similar to their contour length, ℓ_c, and can therefore neither be accurately described as rigid rods nor as flexible chains. The elasticity of a semiflexible polymer is entropic, as that of a flexible polymer, because it has many more possibilities to be curved than to be straight. Upon stretching, the number of available conformations and therefore the entropy is reduced, which generates an opposing force. The energy per unit length of the chain depends on the bending of the chain and the work against the applied force f,

$$H = \frac{\kappa}{2}\left(\nabla^2 u\right)^2 + \frac{f}{2}\left(\nabla u\right)^2, \qquad (4.1.4)$$

with bending modulus κ and deviation $u(x)$ of the chain from the straight conformation along the x axis [MacKintosh 95].

Using the equipartition theorem, the average end-to-end length $L(f, \ell_c)$ can be determined from the transverse thermal fluctuations of u at a certain force f:

$$L(f, \ell_c) - L(0, \ell_c) = \frac{\ell_c^2}{\ell_p \pi^2} \sum_{n=1}^{\infty} \frac{f/f_c}{n^2(n^2 + f/f_c)}, \qquad (4.1.5)$$

with a scaling force $f_c = \kappa \pi^2/\ell_c^2$. The equilibrium end-to-end length of a filament with contour length ℓ_c is given by $L(0, \ell_c) = \ell_c(1 - \ell_c/6\ell_p)$ [Storm 05].

The force diverges as $(\ell_c - \ell)^{-2}$ as the length approaches the contour length, $\ell \to \ell_c$

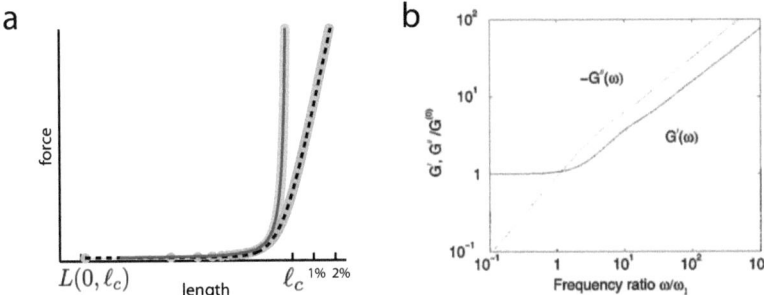

Figure 4.2: a) Force-extension curve for a semiflexible polymer according to the Worm-like Chain (WLC) model. For the purely entropic elasticity, the force-length curve starts from zero at the equilibrium length $L(0, \ell_c)$ and diverges as the length approaches the contour length, ℓ_c. Grey: exact solution eq. (4.1.5), solid red line: interpolation formula (4.2.1). With an additional stretch modulus, the chain can be stretched beyond its contour length up to small strains of 1-2%. Blue: exact solution from [Storm 05], dashed black line: interpolation formula (4.2.2). b) Real part G' (solid line) and absolute value G'' of the imaginary part (dotted line) of the frequency-dependent shear modulus according to eq. (4.1.8) relative to the plateau modulus $G^{(0)}$. From [Gittes 98].

(see Fig. 4.2a), leading to a differential elastic modulus of the chain $K = df/d\ell$ that increases as $f^{3/2}$ in the nonlinear regime. The elastic modulus G of an isotropic network of semiflexible filaments with polymer density ρ is given by $G = \frac{1}{15}\rho\ell_c K$ [Gittes 98]. An additional enthalpic stretching modulus can be included in the force-length relationship, eq. (4.1.5). The resulting theoretical shear modulus accurately describes the static elastic behavior of a wide variety of reconstituted in-vitro biopolymer networks [Storm 05].

The dynamic frequency-dependent properties of such networks can be obtained by including friction against a viscous background medium with viscosity η, resulting in a transverse drag coefficient ζ [Gittes 98]. The thermal fluctuations along the contour length of the filament relax with a spectrum of time constants,

$$\tau_n = \tau_\ell/(n^4 + n^2 f/f_\ell), \qquad (4.1.6)$$

with a characteristic relaxation time $\tau_\ell = \zeta \ell_c^4 / \kappa \pi^4$ and Euler force $f_\ell = \kappa \pi^2 / \ell_c^2$ of the longest mode, and f the force along the filament backbone. The frequency-dependent susceptibility of the filament can be derived using correlation analysis:

$$\alpha(\omega) = \frac{\ell_c^3}{k_B T \ell_p \pi^4} \sum_{n=1}^{\infty} \frac{1}{(n^4 + n^2 f/f_\ell)(1 + i\omega\tau_n)}. \qquad (4.1.7)$$

The frequency-dependent shear modulus $G(\omega)$ of the corresponding isotropic network of semiflexible polymers including relaxation is then given by

$$G(\omega) = \frac{1}{15}\rho\ell_c/\alpha(\omega) - i\omega\eta. \qquad (4.1.8)$$

At low frequencies, this becomes a frequency-independent plateau modulus, whereas at high frequencies, the sum in eq. (4.1.7) is replaced by an integral, resulting in a power-law increase of the shear modulus G with frequency, $G(\omega) \propto \omega^{3/4}$, as shown in Fig. 4.2b.

While this model accurately describes the elastic as well as dynamic properties of actin and other semiflexible polymer networks [Gardel 04, Storm 05], it fails to account for the mechanical properties of cells in its present form. On the one hand, a quantitative derivation of the shear modulus is difficult since structural parameters such as mesh size or filament length are not easily accessible in cells. On the other hand, the dynamic properties of cells show much smaller power law exponents at intermediate frequencies. Only at frequencies above 100 Hz, this weak power-law regime crosses over to the $\omega^{3/4}$ scaling behavior that is predicted by the WLC model [Deng 06].

It can therefore be concluded that cytoplasmic viscosity as origin of the dissipative properties of the cytoskeleton only comes into play at high frequencies, whereas at low to intermediate frequencies, unbinding of weak attractive bonds is the dominating mechanism of energy dissipation. Such weak interactions, which could be filament entanglements or dynamic crosslinks in the case of the cytoskeleton, slow down the relaxation of a wormlike chain against its viscous surroundings. An extension to the classical WLC model that takes into account such effects is called "Glassy Wormlike Chain" (GWLC). It correctly predicts the long-time relaxation dynamics of entangled actin networks [Semmrich 07]. The feasibility of this approach for describing the experimental results of this work will be discussed towards the end of this chapter.

Even in the absence of any entropic filament fluctuations, the stochastic rupture of many parallel weak bonds under force can lead to characteristic stress relaxation dy-

4.1 Introduction

namics. This mechanism certainly plays a role for the rheological properties of living cells, as they interact mechanically with their environment via highly dynamic complexes of weak protein bonds, called focal adhesions. In the following section, the canonical description of such force-enhanced unbinding of biological bonds is introduced.

4.1.4 Force-induced unbinding of biological bonds

Internal cell structure and adhesion are goverend by weak, noncovalent bonds that are constantly turning over. The lifetime of such biochemical bonds under force depends on the rate of force application and duration of loading. The formalism to describe the role of mechanical force for the rate of bond dissociation was first introduced by Bell [Bell 78] as an extension of transition state theory and will be reviewed in the following.

In the absence of force, the rate of spontaneous thermal unbinding of a weak attractive bond at temperature T is equivalent to the likelihood of crossing the binding energy barrier E_b, expressed by the Arrhenius equation,

$$k_{\text{off}} = r_0 e^{-E_b/k_B T}, \tag{4.1.9}$$

with an attempt frequency r_0. Under application of an external force f, the energy landscape is tilted by an additional potential $-fx$. For a sufficiently sharp energy barrier, the position x_b of the transition state along the reaction coordinate does not change, and the external force simply lowers the energy barrier by an amount fx_b (Fig. 4.3a). The resulting unbinding rate is then given by

$$k'_{\text{off}} = r_0 e^{-(E_b - fx_b)/k_B T} = k_{\text{off}} e^{f/f_T}, \tag{4.1.10}$$

with a critical thermal force $f_T = k_B T / x_b$ above which the unbinding rate rises exponentially with force.

The original concept by Bell has been generalized and refined by Evans & Ritchie [Evans 97], and has proven to accurately describe the dynamics of bond dissociation under constant load for individual [Merkel 99] and multiple parallel biomolecular bonds [Seifert 00, Prechtel 02]. Dynamic rebinding has been included in an extended cluster model [Erdmann 04] to study turnover and adhesion dynamics.

As has already been pointed out above, the dissipative mechanical properties of cells at timescales longer than milliseconds are probably not dominated by cytoplasmic viscosity, but by elastic energy release due to unbinding of weak interactions. The force-induced

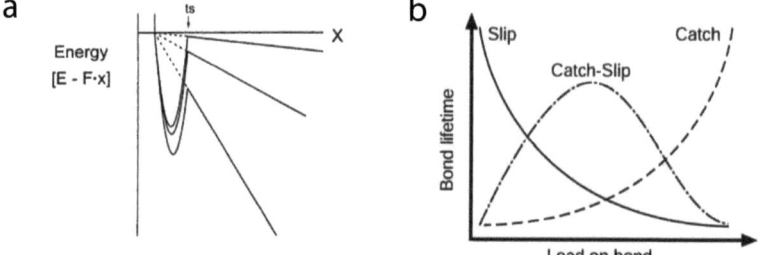

Figure 4.3: a) Conceptual view of the energy landscape of a biological bond along a reaction pathway in unbonding, from [Evans 97]. Application of force to a bond tilts the energy landscape. For a deep harmonic well, the transition state (ts) remains fixed. b) Illustration of the responses of catch, slip, and catch-slip bonds to imposed loads when measured in terms of bond lifetime, from [Guo 06]. For a catch-slip bond, the lifetime first increases when force is applied, reaches a maximum, and then decreases again. Catch bonds represent a possible mechanism to explain an increase of bond lifetimes and hence more elastic behavior with increasing force or prestress.

4.1 Introduction

hopping of elements across energy barriers in the SGR framework is conceptually identical to the distortion of the unbinding energy landscape as described in the Bell model.

One distinct feature that makes cells stand out from all other soft materials such as colloids or polymer networks has not been adressed by any of the models mentioned so far: the ability to actively generate forces. As shown in the experimental results, the active acto-myosin generated prestress of the cytoskeleton plays a crucial role for the mechanical properties of cells. In the following section, the sliding filament model of acto-myosin force generation in muscle will be reviewed, which, in a generalized form, constitutes the basis for the model of cell mechanics developed in this work.

4.1.5 The Sliding Filament (SF) model

Experimental evidence suggests that nearly all aspects of mechanical cell behavior are closely associated with the cells contractile machinery of actin and myosin filaments. The molecular mechanism of contractile force generation in most eukaryotic cells is the same as in skeletal and smooth muscle: the interaction between actin filaments and myosin(II) molecular motors. Nowadays, the details of this mechanism are known down to the molecular level, but the corresponding mathematical model of muscle, called sliding filament theory, was already introduced in 1957 [Huxley 57], and is still used. It captures all relevant features of force generation and length adaptation in skeletal muscle and, within limits, also of smooth muscle and contractile non-muscle cells [Fredberg 96, Mijailovich 00, Fernandez 08].

A schematic representation of the sliding filament model is shown in Fig. 4.4a. The myosin filament is fixed in space, and the actin filament can be displaced in x-direction, where positive x corresponds to muscle lengthening. Crossbridges can attach to the actin filament within a range $0 < x < h$ with a position-dependent probability $f(x) = f_1 x/h$. Detachment of crossbridges occurs with a probability $g(x) = g_1 x/h$ at positive x, and a constant high probability g_2 at negative x (Fig. 4.4b). Crossbridges behave as Hookean springs and develop a tension $F = kx$ under elongation x. Attachment of myosin crossbridges at $x > 0$ therefore occurs in a force-generating state.

Let $n(x)$ be the probability that an arbitrarily chosen crossbridge with displacement x is attached. Crossbridge cycling follows a first-order kinetic rate equation:

$$\frac{dn(x)}{dt} = [1 - n(x)]f(x) - n(x)g(x). \quad (4.1.11)$$

Under the condition of muscle shortening at constant speed v, the solution of the rate

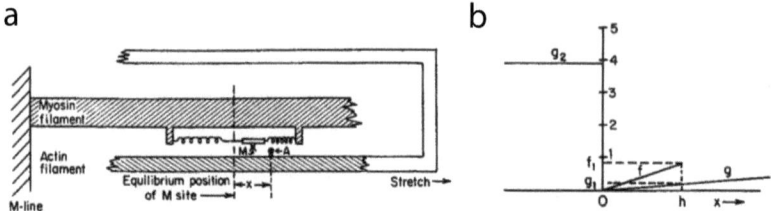

Figure 4.4: a) Huxley's 1957 model for force generation of striated muscle. The thick filament is fixed at the M-line; the thin filaments move to the right if the muscle is stretched. b) Rate constants for attachment and detachment are not symmetric around $x = 0$. The unit of the vertical scale is the value of $(f + g)$ at $x = h$. From [Huxley 57].

equation for the region $0 < x < h$ is

$$n(x) = \frac{f_1}{f_1 + g_1} \left[1 - e^{-(f_1 + g_1)(h^2 - x^2)/2hv} \right]. \qquad (4.1.12)$$

In a similar way, the developed tension, rate of energy liberation, and Hill's force-velocity equation can be derived [McMahon 84].

The sliding filament model is still used to describe the mechanics of skeletal and smooth muscle – only the number of possible bound and unbound states of actin and myosin in newer versions of the model has increased to reflect the increasing knowledge about the molecular details of crossbridge cycling. For instance, the change of hysteresivity or energy dissipation in smooth muscle can be explained by a stronger binding of myosin to actin in the absence of ADP during steady-state contraction, the so-called *latch* state [Hai 88].

The elastic and dissipative mechanical properties of cells are directly linked to their contractile prestress, as has been shown in the previous chapter. Although skeletal muscle has a highly organized, almost crystalline structure compared to the disordered cytoskeleton of most eukaryotic cells, the molecular mechanism of force generation is identical. It seems obvious to incorporate the basic formalism of the sliding filament model into a comprehensive model of cell mechanics.

4.1.6 Outline for a combined generalized model

The main challenge in developing a model is finding the right balance between complexity and simplicity. In the case of cell mechanics, a certain amount of detail is necessary in order to reflect our knowledge about the internal structure of the cell and to explain the diversity of experimental observations. At the same time, a model of cell mechanics should be simple enough to allow for intuitive understanding. It should have as few parameters as possible in order to make useful predictions given the limited control of boundary conditions in biological experiments.

The traditional approach towards modeling of viscoelastic properties are mechanical equivalent circuits composed of springs and dashpots. Every elastic or viscous compartment of the real system is represented by the spring constant of a Hookean spring or the friction coefficient of an ideal dashpot. Arrangements of springs and dashpots give rise to stress relaxation with a few characteristic time constants. Until recently, this approach from classical mechanics has been the canonical way to fit data from cell rheology experiments [Yagi 61, Bausch 98, Feneberg 04].

Experimental improvements of time resolution, however, have revealed a constant ratio of storage and loss moduli in cells [Fabry 01, Hoffman 06]. Such behavior, known as "structural damping" in architecture since the 1920s, cannot be modeled efficiently using separate elastic and viscous elements, but requires instead that energy storage and dissipation are structurally coupled. Consequently, there must be a broad distribution of diverse disordered elements that store and dissipate mechanical energy at different time scales, consistent with the disordered structure of most biological materials. This rules out the classical spring-dashpot equivalent circuit.

The following proposal for a model is in parts phenomenological, as it embodies abstract concepts of nonlinear elasticity and soft glassy rheology, and in parts mechanistic when it comes to filament elasticity, bond dissociation, and force generation, which all have structural counterparts in the cell. It should be seen as an equivalent circuit that incorporates sufficient microscopic detail to link physical properties to biological functions, and to demonstrate that the mechanical behavior observed in cells and other biological materials originates from generic structural properties, independent of the actual molecular constituents of the system.

As a summary of this introduction and in anticipation of the more detailed description that will follow in the next section, the essential features of the proposed mechanical model are outlined:

- Stress is carried by many elastic elements in parallel which represent individual cytoskeletal filaments or crosslinker proteins.

- Individual elements behave as entropic springs with an additional stretch modulus, resulting in force-elongation curves as described by the WLC model.

- Initially, elements are randomly oriented with respect to external force to account for the geometric disorder of the cytoskeleton.

- Elements yield with force-dependent rates according to the Bell model, giving rise to an effective viscosity. All other sources of viscosity are neglected.

- Lifetimes of elements are broadly distributed to reflect structural heterogeneity, leading to power-law rheology as in soft glassy materials.

- New elements can attach in a force-generating state, analogous to the sliding filament model of Huxley, to reflect acto-myosin force generation.

4.2 Geometry and stress-strain relationship

The following section contains a detailed description of the stucture and elastic properties of the model. The geometry is described, and differences to the original sliding filament geometry are discussed. Nonlinear elastic properties arise from three different mechanisms: the nonlinear entropic elasticity of individual elements, the random orientation and geometric alignment of elements, and the prestress of elements. The static stress-strain curve and stress-dependent stiffness of the model are calculated numerically.

4.2.1 Sliding filament geometry

The material is represented by an uniaxial arrangement of two stiff filaments that are crosslinked by parallel elastic elements, as shown in Fig. 4.5. This structure is similar to the sliding filament model of muscle [Huxley 57], with the difference that the crosslinks in the model do not necessarily correspond to individual myosin motor molecules. Instead, they stand for any kind of mechanically loaded elastic cytoskeletal elements, such as actin or intermediate filaments, flexible crosslinks, or adhesion bonds. This generalized sliding filament geometry should be understood as a mechanical equivalent circuit of the cytoskeleton.

Another important difference to the sliding filament model of muscle is that the stiff upper and lower filaments do not represent actual sarcomeric structures, but merely serve as fixed mechanical endpoints. They correspond, e.g., to a bead bound to the apical surface of the cell and the stiff substrate the cell is attached to, or the upper and lower plates of a microplate rheometer.

In most rheological experiments, stress and strain are applied and measured along the same direction, and the resulting experimental data are one-dimensional. This is reflected in the effectively one-dimensional geometry of the model. The idea behind this abstraction is that topological details are not relevant for the generic mechanical properties of the cell.

4.2.2 Nonlinear elasticity

Three mechanisms contribute to the nonlinear elastic behavior of the model: the nonlinear stress-strain relationship of individual elements, the geometric disorder of the system, and the prestress of elements upon attachment.

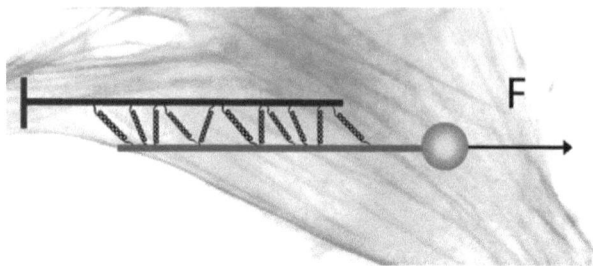

Figure 4.5: Illustration of the geometry of the generalized sliding filament model. The material is represented by a mechanical equivalent circuit that consists of an uniaxial arrangement of two stiff filaments, crosslinked by parallel elastic elements. The two filaments represent the fixed endpoints of the geometry, i.e., in a microrheology experiment, they would correspond to the bead and the rigid substrate. The elastic elements stand for any kind of mechanically loaded elastic cytoskeletal elements, such as actin or intermediate filaments, flexible crosslinks, or even adhesion molecules.

Nonlinear elasticity of individual elements

The nonlinear entropic elasticity of individual filaments contributes to the stress stiffening of networks of semiflexible filaments such as the cytoskeleton [Storm 05, Onck 05]. The elastic elements are modeled as wormlike chains with entropic elasticity in series with an additional enthalpic stretching modulus, as described in [Storm 05].

The stress-strain curve of a real semiflexible filament can be divided into three regimes (Fig. 4.2): a small-strain regime starting from the equilibrium force-free configuration, where fluctuations can be pulled straight with little effort; a nonlinear regime as the length approaches the contour length; and a linear enthalpic stretching regime as molecular bonds in the chain are stretched and the filament is elongated beyond its contour length before it ruptures.

For reasons of computational efficiency, an interpolation of the exact WLC solution is used for the force-elongation curves of individual elements in the model. The nonlinear regime of the entropic contribution to WLC elasticity follows the inverse quadratic formula described in [Bustamante 94]:

$$f = \frac{k_B T}{\ell_p} \left[0.25 \left(1 - \frac{L}{L_0}\right)^{-2} - 0.25 + \frac{L}{L_0} \right]. \tag{4.2.1}$$

4.2 Geometry and stress-strain relationship

For the combined entropic and enthalpic elasticity used in the model, the following interpolation function is used to replace the exact solution described in [Storm 05]:

$$f = \frac{k_B T}{\ell_p} \left[a_1 \ln \left(1 + e^{a_2(L/L_0 + a_3)} \right) \right] \qquad (4.2.2)$$

with $a_1 = 5.404 \cdot 10^3$, $a_2 = 195.9$ and $a_3 = 0.9941$. The two interpolation formulas are compared to the real curves in Fig. 4.2a.

Nonlinearity due to geometrical disorder and alignment

As pointed out above, the sliding filament geometry of the model does not necessarily depict ordered muscle-like cellular structures such as stress fibers. In general, the cytoskeletal structures that are represented by the parallel arrangement of elastic elements are randomly oriented with respect to the applied force. This geometrical disorder is modeled by a distribution of attachment angles of the elements in the initial state (Fig. 4.6a). Attachment angles are uniformly distributed between ±45°.

As the system is strained and the two "filaments" slide past each other, the elements start to align in the direction of strain. With increasing strain, more and more elements contribute to the elastic resistance of the material. This mechanism accounts for fiber alignment due to strain as one origin of stress stiffening in filamentous materials [Fung 93].

Nonlinearity due to prestress-stiffening

The third mechanism that is responsible for the dependence of stiffness on stress is due to the prestress of elements. The resistance of an elastic element to deformation perpendicular to its axis is proportional to the already present tension along its axis. This "guitar string effect" of fiber stiffness also motivated the application of the tensegrity concept known in architecture to the cytoskeleton [Ingber 03]. In the generalized sliding filament geometry, this stiffening mechanism is accounted for: the force required to move the two stiff filaments against each other depends on the resistance of the elastic crosslinks against lateral deformation and, hence, on their prestress.

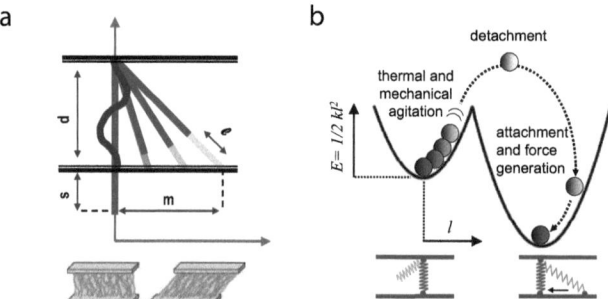

Figure 4.6: a) Geometrical disorder is modeled by a distribution of attachment angles of the elements in the initial state. As the system is strained and the two "filaments" slide past each other, the elements start to align in the direction of strain. This mechanism accounts for fiber alignment due to strain as one origin of stress stiffening in filamentous materials [Fung 93]. b) New elements can appear in an already strained conformation to describe actomyosin force generation. In the potential well picture of SGR, attachment at a force-generating position corresponds to new elements that start not at the bottom of a trap, but at some position to the right or to the left. Thus, the model describes a kind of "active soft glassy rheology".

4.2.3 Stress-strain curve: numerical results

The initial configuration of the model is a sliding filament-like arrangement of N parallel elements. The initial attachment angles are uniformly distributed between -45° and +45° to represent geometrical disorder. The force-elongation curves $f(l/\ell_c)$ of the elements are calculated using the interpolation formula eq. (4.2.2) for WLC elasticity, as shown in Fig. 4.2a. The contour length ℓ_c of all elements is set so that the restoring force f in the initial elongation l/ℓ_c is equal to a predefined prestress force, f_p.

The static elastic stress-strain curve in the absence of yielding is calculated by setting a global strain corresponding to a displacement of the two filaments against each other. For each strain, the elongation of each element and the corresponding restoring force is calculated. All individual forces are summed up to a global system force, or stress. The resulting stress-strain curves for different values of the prestress force f_p are shown in Fig. 4.7a. The corresponding stiffness-vs.-stress curves are obtained by taking the numerical derivative of the stress-strain curves and plotting them against stress (Fig. 4.7b). The result closely resembles the stress stiffening observed experimentally in cells, as described in section 3.3: the stiffness at small forces and the amount of stiffening at larger forces both depend on prestress.

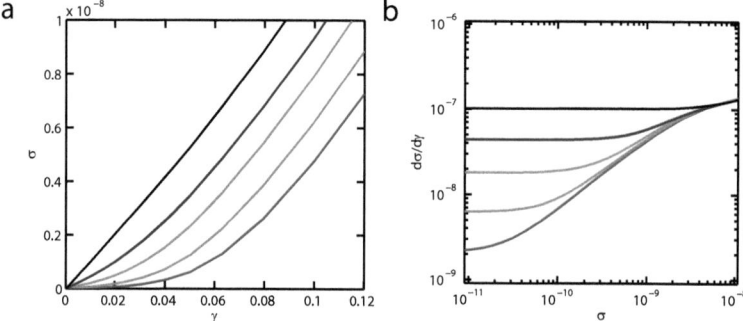

Figure 4.7: a) Stress σ vs. strain γ for different prestress in the static case without unbinding. The curves from bottom to top correspond to prestress values of $f_p/f_c = 0.001, 0.003, 0.01, 0.03$ and 0.1. Larger values of prestress correspond to more linear stress-strain curves. The unit of stress has no physical meaning here since it depends on the experimental geometry. b) Differential stiffness $d\sigma/d\gamma$ vs. stress σ for the same data as shown in a). The amount of nonlinear stress stiffening depends on the value of prestress. This figure closely resembles the experimental results of creep measurements shown in Fig. 3.11.

4.3 Time-dependent behavior

In this section, the mechanism of stress relaxation due to element unbinding is explained. The distribution of yield lengths and lifetimes are explained from the disordered structure of the system. The force-dependence of the element lifetimes in the case of classical slip bonds and for catch-slip bonds is derived, and differences to the SGR model are discussed. Finally, the phenomenological power law exponents for creep and stress relaxation are calculated numerically. The actual creep response of the model is simulated using Monte Carlo methods and compared to the experimental results.

4.3.1 Stress relaxation due to spontaneous unbinding

The dominating mechanism of stress relaxation in the cytoskeleton at timescales above a few milliseconds is the unbinding of weak non-covalent bonds. Bond lifetime without external force can be described as thermal activation over an energy barrier E_b: $\tau = \tau_0 \exp(E_b/k_B T)$, where k_B is the Boltzmann factor and T the temperature, and τ_0 some intrinsic timescale of unbinding. In the model, each elastic element has a characteristic lifetime, corresponding to a certain height of the energy barrier for unbinding.

Distribution of bond lengths and yield energies

Following the description of [Bell 78] for the unbinding of weak biological bonds, the yield energy E_b is associated with an unbinding force f_c which is given by $f_c = E_b/x_b$, where x_b is the coordinate of the energy barrier in direction of force, or the yield elongation of the bond. In the model, f_c is identical for all elements since the elastic force for the WLC-like elements depends only on their *relative* extension l/ℓ_c (eq. (4.2.2)), and all elements can sustain the same *relative* extension or strain before they break. The *absolute* yield length x_b, however, is proportional to the absolute element length which can be broadly distributed in a heterogenous network such as the cytoskeleton. In polymerized actin networks [Edelstein-Keshet 98, Xu 99] and simulations thereof [Heussinger 07], filament lengths were found to be exponentially distributed. Consequently, the distribution of possible yield lengths in the model is also exponential: $\rho(x_b) \sim e^{-x_b}$. In our case of constant f_c, the yield energies $E_b = f_c x_b$ follow the same distribution as the yield lengths x_b.

Bond lifetimes and power law rheology

The resulting exponential distribution of yield energies can be written as $\rho(E) \propto e^{-E/E_g}$, where the energy scale is defined by E_g. Cytoskeletal bonds such as actin-myosin or integrin-ligand interactions unbind spontaneously with typical lifetimes of a few seconds. In this case, the typical energy E_g must be smaller than $k_B T$ so that thermal fluctuations can overcome the energy barrier for unbinding.

This description is formally identical to the trap model of soft glassy materials described by Bouchaud [Bouchaud 92, Sollich 97], only the thermodynamic temperature $k_B T$ is used instead of the noise temperature. If the distribution of element energies is written as $\rho(E) \propto e^{-xE/k_B T}$, then the dimensionless parameter $x = k_B T / E_g$ takes the same role as the noise temperature in the SGR model. It denotes the factor by which the typical yield energy exceeds the thermal energy, $k_B T$. As long as $x > 1$, spontaneous thermal unbinding takes place, and the system evolves towards the equilibrium Boltzmann distribution of energies,

$$P_{eq}(E') = \rho(E') e^{E'/k_B T} = e^{-\mathcal{E}} e^{\mathcal{E}/x}, \qquad (4.3.1)$$

where $\mathcal{E} = E'/E_g$.

As in the SGR model, such a system exhibits power law rheology with an exponent $\beta = x - 1$, e.g. $G(\omega) \propto \omega^\beta$, as long as $1 < x < 2$. In contrast, if $x \to 1$, all yield energies have the same probability and P_{eq} ceases to be normalizable. This corresponds to a glass transition at temperature T_g where typical bond energies $E_g = k_B T_g$ are equal to or larger than the thermal energy.

4.3.2 Force-dependent lifetimes

Soft Glassy Rheology and active prestress

The most important difference to the SGR model apart from the temperature concept is the mechanism of active force generation: analogous to the sliding filament model, new elements can appear in an already strained conformation to describe acto-myosin force generation. In the sliding filament geometry shown in Fig. 4.4, this corresponds to attachment at a position slightly off the equilibrium position. In the trap picture of SGR, attachment at a force-generating position corresponds to new elements that start not at the center of their potential well, but at some position to the right or to the left, as depicted in Fig. 4.6b. In this "active soft glassy rheology" picture, new bonds experience

4.3 Time-dependent behavior

a typical force f_p upon attachment that is proportional to the amount of contractile prestress.

The rate at which new bonds appear is constant and independent of the detachment rate or of the external force. In the model, the attachment rate is set so that in steady state, attachment and detachment rates are roughly the same, similar to the SGR model where a new element appears for every old one that yields.

Force induced unbinding and fluidization

If a bond experiences a force f, its lifetime is reduced as described by the Bell formalism [Bell 78]: $\tau = \tau_0 e^{(E_b - f x_b)/k_B T}$. Under prestress, new bonds already carry a force f_p, which reduces their lifetime accordingly. If f_p is expressed as a fraction of the total yielding force, $f_p = \alpha f_c$ ($\alpha < 1$), the bond lifetime under prestress becomes

$$\tau(E) = \tau_0 e^{E_b(1-\alpha)/k_B T}. \tag{4.3.2}$$

As in the force-free state, the steady-state probability $P(E)$ for finding a potential well of depth E is proportional to the corresponding bond lifetime:

$$P_{eq}(E') = \rho(E') e^{E'(1-\alpha)/k_B T} = e^{-\mathcal{E}} e^{\mathcal{E}/x'} \tag{4.3.3}$$

with $\mathcal{E} = E'/E_q$ and $x' = x/(1-\alpha)$. This means that in the model, the detachment rate and rheological power law exponent increase with increasing f_p, resulting in more fluid-like behavior under higher prestress.

This is in contrast to the actual behavior observed in living cells: the higher the actomyosin-generated prestress, the lower the rheological power law exponent or, in other words, more contractile cells exhibit more solid-like mechanical response. A reduction of bond lifetime by force cannot account for this observation.

Increase of bond lifetimes with prestress

In order to describe the experimentally observed decrease of β or, correspondingly, increase of bond lifetimes with prestress, the unbinding rate of elements is written as the sum of two Bell processes, one of which has a negative dependence of the dissociation rate on the reaction coordinate. The mean bond lifetime $\bar{\tau}$ is the inverse of the sum of the reaction rates for the two possible unbinding reactions,

$$1/\bar{\tau} = k_0 e^{(E_c - f x_b)/k_B T} + k_0' e^{(E_c + f x_b')/k_B T} \qquad (4.3.4)$$

with two different intrinsic rate constants k_0, k_0' and two bond lengths x_b, x_b'.

Although this mathematical description closely follows the model of catch-slip behavior (Fig. 4.3) proposed by [Pereverzev 05], it is meant in a more generic sense. Three possible mechanisms behind an increase of bond lifetimes with increasing prestress are given in the following.

Catch-slip bonds, the latch hypothesis, and active remodeling

Several protein-protein interactions have been reported to show a catch-slip transition where bond lifetime first increases with increasing force ("catch" mechanism) up to a certain maximum lifetime at a characteristic force, and then decreases again as force is increased further ("slip" behavior), as in the conventional Bell picture. Such behavior has specifically been described for acto-myosin [Guo 06] and integrin-ligand bonds [Kong 09], both of which are important for the mechanical properties of the cytoskeleton. The mechanistic explanation behind such counter-intuitive characteristics could be, for example, a hook-like bond structure that dissociates spontaneously when subjected to thermal fluctuations, but gets stuck if a force is applied.

In the case of actomyosin, an alternative explanation is a "latch" mechanism as it has been described for skeletal and smooth muscle. In smooth muscle cells, the more solid-like behavior under high prestress arises from reduced actomyosin cycling and energy dissipation due to a strongly bound "latch" state of actin and myosin [Fredberg 96]. This could be accounted for by the model via introduction of a second bound state, and additional transition rates between the bound states. It has, however, been shown in [Pereverzev 05] that such a two-state description is analogous to the simpler two-pathway unbinding formalism of eq. (4.3.4).

A more general explanation is that cells under different prestress have a completely different structure, giving rise to different distributions of element lifetimes and bond strengths. On the one hand, such active remodeling certainly takes place in living cells and could be accounted for by extending the model correspondingly. On the other hand, cells modulate their prestress and hence mechanical properties very rapidly within seconds, whereas remodeling takes places on timescales of minutes to hours. It therefore makes sense to treat the two factors, prestress and remodeling, separately.

4.3.3 Numerical results

Steady-state power law exponent

In the absence of flow, the system approaches a steady state where the total yielding rate equals the attachment rate, and the number of bonds remains roughly constant. This remains valid if new bonds attach with a prestress force f_p. The probability to find a bond of a certain yield energy is proportional to the lifetime of this bond and the density of states of possible bonds. For the two-pathway mechanism (eq. (4.3.4)) and with prestress, it is given by

$$P_{eq}(E') = \rho(E') \left[k_0 e^{(E_c - f x_b)/k_B T} + k'_0 e^{(E_c - f x'_b)/k_B T} \right]^{-1}. \quad (4.3.5)$$

A steady state distribution of yield energies corresponds to a steady-state lifetime distribution. In the case of exponentially distributed energies or bond lengths and force-independent lifetimes, the steady-state lifetimes are power-law distributed, as shown in [Sollich 97]. For the two-pathway formalism, the distribution is more complex, and eq. (4.3.5) was therefore solved numerically to derive the lifetime distributions and corresponding rheological power law exponents.

Ensembles of yield energies following the distribution given in eq. (4.3.5) were generated from uniformly distributed pseudo-random numbers using rejection sampling. The corresponding lifetimes τ were calculated by inserting these energies into eq. (4.3.4). The resulting τ distributions are shown in Fig. 4.8a. These distributions were fit to a power law in order to characterize the steady-state rheological response to small forces under different amounts of prestress.

The result is shown in Fig. 4.8b. The dependence of the power law exponent β on force is biphasic and exhibits a minimum at about 10% of the critical force. The interpretation in the cytoskeletal context is that the minimum of β corresponds to the highest possible level of steady-state prestress in the cell. If the force on elements due to internal prestress would be beyond 10%, the unbinding rate would increase again with force, leading to a further increase of force onto the remaining elements upon unbinding and eventually to rupture of all elements.

If an external force is applied to the system while it is in a steady state somewhere on the descending part of the force-β-curve in Fig. 4.8b, β will follow this curve and first decrease, then increase with increasing force. Experimentally, however, a decrease of β with increasing external force is rarely, if ever, observed (see section 3.2), even though the occasional rupture of beads indicates that the applied external force per element

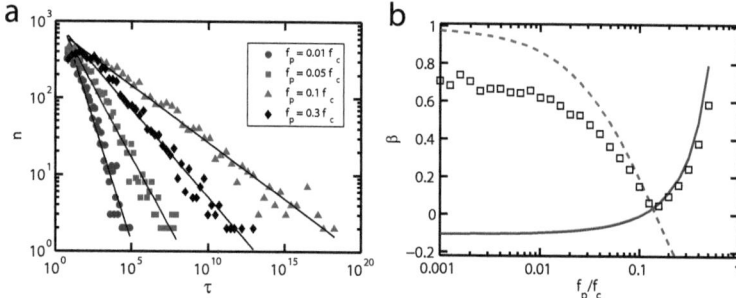

Figure 4.8: a) Histogram of element lifetimes τ under different prestress, for a total number of n=5000 elements, according to eq. (4.3.4). The solid black lines are fits to a power law. b) Black squares: result of the power law fit to the lifetime histograms. The y-axis (β) is the negative of the power law exponent from the fit and corresponds to the rheological power law exponent of the linear viscoelastic response of the system. The solid blue line corresponds to the result for pure slip-bond behavior. In regions where $\beta<0$, no spontaneous thermal unbinding of bonds can take place. The dashed red line shows β for pure catch-bond behavior. Spontaneous unbinding ($\beta>0$) via the catch pathway only takes place at low forces but is inhibited by large forces ($\beta<0$).

must be a substantial fraction of the critical force, f_c. This is a weak point of the model and can possibly be accounted for in the future by relaxing the assumption of identical yield force and critical strain for all elements, thereby "washing out" the characteristics of the force-β-curve under external force.

Creep and flow lead to reaction rates that change over time since the external force is redistributed among the remaining elements with every unbinding event. Moreover, the increased unbinding rates due to external force lead to an imbalance of unbinding and rebinding rates and a change of the total number of elements. In order to determine the properties of the model in the absence of a steady state, the creep response was simulated numerically.

Monte Carlo simulations of creep

The creep response of the model was determined by Monte Carlo simulations. The initial exponentially distributed bond lengths or energies were generated from uniformly

4.3 Time-dependent behavior

distributed random numbers using the method of inverse transform sampling, or probability integral transform. It states that if r is a random variable with a standard uniform distribution on the interval $[0, 1]$, then the random variable $s = F^{-1}(r)$ has the distribution described by F. In the case of an exponential distribution,

$$s = -\frac{\ln r}{\lambda} \qquad (4.3.6)$$

are exponentially distributed with $\mathrm{pdf}(s) \sim e^{-\lambda r}$.

The numerical implementation of the model was done in MATLAB. During a simulation run, the following steps were carried out once for every timestep:

1. Starting from the preset global force $F(t_n)$, the global length $L(t_n)$ is calculated using a force balance: the difference between the global force and the sum of all individual element forces is minimized using the fzero function of MATLAB, which is based on the minimization algorithm described in [Brent 73].
2. The individual elongations $l^i(t_n)$ and forces $f^i(t_n)$ for all elements are calculated from the new global length $L(t_n)$ that was derived in the previous step.
3. From the individual element forces, the unbinding rates k_{off}^i for all elements are calculated using a superposition of two Bell processes (eq. (4.3.4)).
4. The reaction that occurs in the current timestep (the element to yield or attach) and the duration of the timestep Δt_n is determined in a Monte Carlo step as described below.
5. If an element detaches, it is removed from the ensemble. If a new element attaches, it is initialized at the current position with prestress force f_p and corresponding contour length ℓ_c, and a new attachment angle is drawn from a uniform distribution.

In step 4 of the procedure described above, the reaction to be carried out and the length of the timestep in which it takes place are determined using the Kinetic Monte Carlo (KMC) algorithm described by Gillespie [Gillespie 77]. First, a vector of cumulative sums

$$R_i = \sum_{j=1}^{i} r_j \quad (i = 1...N) \qquad (4.3.7)$$

of the rates r of all possible reactions is calculated. This includes the force-induced detachment of any element as well as the attachment of a new element that takes place at a constant force-independent rate. A uniform random number $u_1 \in [0, 1]$ is drawn, and the reaction that actually takes place is determined by finding the i for which $R_{i-1} < u_1 R_N \leq R_i$.

The possible timestep lengths are exponentially distributed, weighted by the sum of all rates. This distribution corresponds to the exponential decay of the probability that no reaction has occured. To determine the length of the current timestep, a second uniform random number $u_2 \in [0, 1]$ is drawn and the timestep is calculated as

$$\Delta t = -\frac{\ln u_2}{R_N}. \tag{4.3.8}$$

The results of simulated step-creep experiments for different prestress are shown in Fig. 4.9. As expected, the creep response closely follows a power law (Fig. 4.9b). The stiffness as well as the amount of stress stiffening depend on prestress (Fig. 4.9c), although the amount of stiffening is not as pronounced as observed in experiments (see section 3.2). The force dependence of the power law exponent β (Fig. 4.9d) qualitatively reproduces the experiments: the relative increase of β is highest for small initial values of β, whereas for the largest initial β there even is a decrease with increasing force magnitude, which is rarely observed experimentally. The dependence of stiffness (Fig. 4.9e) and power law exponent (Fig. 4.9f) on prestress is in agreement with experiments, as is, in consequence, the relationship between stiffness and β (Fig. 4.9g). A list of all parameters and their values used in the simulations is given in the appendix.

In summary, the model essentially reproduces the experimentally observed time and force dependence of the creep response. In order to obtain a more quantitative agreement between experiments and simulations, the details of the geometry and the unbinding dynamics have to be refined further. In its current state, the model does not allow to derive physically meaningful parameters values by fitting to experimental data. It nevertheless gives an attractive explanation for the structural mechanism behind the mechanical properties of cells.

4.3 Time-dependent behavior

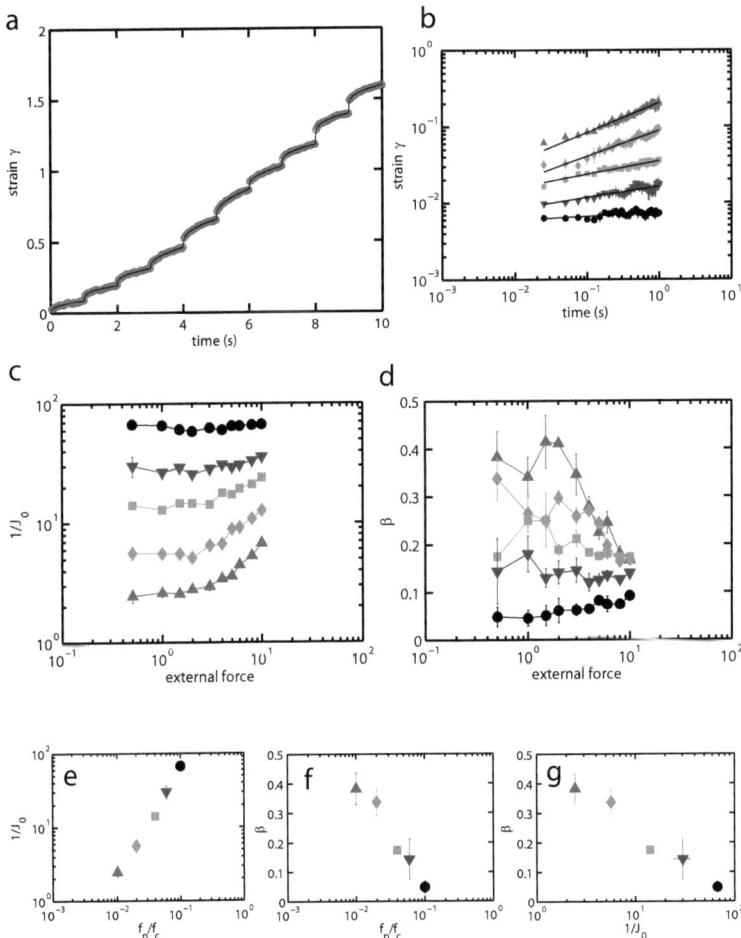

Figure 4.9: a) Example for the simulated creep response of the model. In all following plots, results were averaged over three simulation runs for every prestress. Prestress values f_p/f_c were 0.01 (green triangles), 0.02, 0.04, 0.06, and 0.1 (black circles). Error bars denote standard errors in all cases. b) Creep response to the first force step vs. time. Solid lines are fits to a power law. c) Stiffness and b) power law exponent vs. applied force. e) Stiffness and f) power law exponent vs. prestress, and g) stiffness vs. power law exponent.

5 Discussion

5.1 Limitations and improvements of the experimental setup

5.1.1 Force magnitude

The magnetic tweezers setup described in this work extends the range of forces accessible in bead-based microrheology up to 100 nN. Precise control of the distance between bead and magnet as well as real-time bead tracking at frame rates up to 40 Hz allow for accurate time-resolved microrheology experiments. In some cases, however, it would be desirable to apply even higher forces. In very stiff cells, for instance, the displacement of beads upon application of few tens of nanonewtons is still too small to be sufficiently above the noise level of the bead tracking. Furthermore, quantitative studies of bead binding strength would require to ramp up the force until all beads have detached.

In ongoing studies, different strategies are being investigated to further increase the applicable force magnitude. For instance, beads made of highly pure carbonyl iron have a larger magnetic susceptibility compared to the iron oxide beads used in this work. This should lead to correspondingly higher forces in a magnetic gradient. First calibration experiments with carbonyl iron powder showed that forces of more than 200 nN at 40 μm distance between magnet and bead are possible (Fig. 5.1a). Commercially available carbonyl iron powder, however, has the disadvantage that is is rather polydisperse and has a broad distribution of particle sizes (Fig. 5.1b). Time-consuming filtration is necessary before these particles can be used for microrheology experiments. The feasibility of various filtration techniques is currently under investigation.

Another approach is to use a superparamagnetic core with even higher permeability and saturation flux density than the HyMU80 described in section 2.1. The only commercially available material that fulfils these requirements is amorphous metal, also called "metallic glass". Not only does it have superior magnetic properties, it is also mechanically much harder, which makes it easier to get a stable sharp tip. Due to its amorphous structure, however, metallic glass is very brittle and breaks easily. Besides that, first experiments with a metallic glass needle showed no improvement of the applicable force over HyMU80. Further studies are currently performed to evaluate metallic glass as core material for magnetic tweezers.

5.1.2 Imaging of intracellular structures

In the experiments described in this work, the displacement of cytoskeletally bound beads in response to application of a predefined force was recorded. Force and displacement were converted into stress and strain. The one-dimensional stress and strain

5.1 Limitations and improvements of the experimental setup

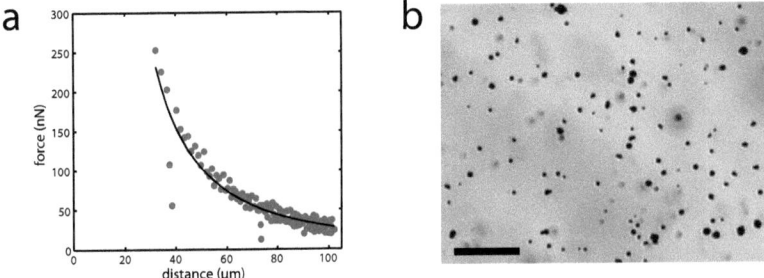

Figure 5.1: a) Force-distance curve for a carbonyl iron bead about 5 μm in diameter. At maximum, forces of 200 nN at 40 μm distance between magnet and bead were measured. b) Carbonyl iron beads in silicon oil under the microscope (magnification 40x). The size distribution is very broad, which makes sorting necessary prior of using them for microrheology. Scale bar = 50 μm.

vectors in the time domain were then used to derive macroscopic rheological parameters of the cytoskeleton. This one-dimensional treatment of stress and strain, although practical, is a gross oversimplification of reality: the cell is not a homogeneous and isotropic material, therefore the actual stress and strain fields are rather complex [Hu 03]. Furthermore, the cytoskeleton responds to the application of external mechanical stress not only by passive deformation, but also by active reorganization and remodeling. The magnetic tweezers setup described in this work can be combined with microscopic imaging of intracellular structures to study these aspects.

In order to image intracellular structures by light microscopy, the organelles or proteins of interest have to be stained selectively by fluorescent dyes in order to distinguish them from the other constituents of the cell. In living cells, this can be done by introducing cDNA into the cell that encodes the protein to be labelled together with a small fluorescent protein such as GFP. The labelled protein is then transiently expressed by the cell for several hours or days. During this time, it can be imaged by illuminating the cell with the excitation wavelength of the fluorescent tag. In contrast to conventional antibody immunofluorescence, this method is suitable for imaging living cells.

Preliminary experiments were performed to demonstrate the possibility of combining confocal live cell imaging and magnetic tweezers microrheology. This work was carried out during two visits to the live cell imaging laboratory of Prof. Clare Waterman at the

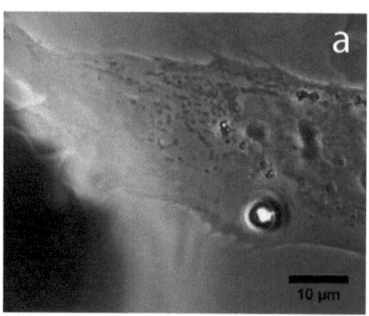 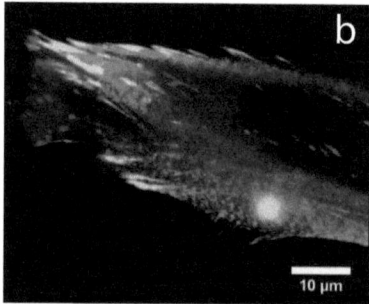

Figure 5.2: a) Living mouse embryonal fibroblast (MEF) under transmitted-light phase contrast illumination. The position of the bound bead and the needle (seen on the bottom left) can be determined from such images, but no information on the intracellular structure is revealed. b) The same cell under illumination with 488 nm and 570 nm incident light. The cell was previously transfected with GFP α-actinin (green) and mCherry Paxillin cDNA (red) to vizualise cytoskeletal structures. Regions where paxillin and α-actinin are both present appear yellow. Stress fibers appear speckled due to their regular α-actinin-rich regions.

Scripps Research Institute in La Jolla, USA.

Actin and α-actinin as strain markers

The most obvious application of fluorescent live cell imaging in cell mechanics is to transfect cells with fluorescently labelled actin in order to image the actin cytoskeleton. The actin network structure then becomes visible, and the stress fibers that serve as mechanical pathways of stress propagation within the cell can be observed. It is, however, difficult to quantify the deformation of these fibers since their intrinsic structure cannot be resolved by staining actin. In order to get quantitative strain field data, it is desirable to image the intrinsic structure of actin stress fibers.

Stress fibers are contractile bundles of actin and myosin filaments that are arranged in a similar manner as in skeletal muscle fibers. They show regular bands of z-disc-like structures that are rich in the crosslinking protein α-actinin. By transfecting cells with GFP-labelled α-actinin, this pattern can be made visible (Fig. 5.2). Deformation and contraction of stress fibers can then be quantified by using the regular α-actinin bands as strain markers. The problem is that large regular stress fibers only appear in very contractile and stiff cells, which require large forces to achieve observable deformation. In first preliminary experiments, the bead was not deflected sufficiently to get any measurable deformation of stress fibers within the measurement time.

Zyxin as an example for active mechanosensing

Cells react to external forces by a large variety of activities such as recruitment of focal adhesion proteins or biochemical signalling events. One example of such behavior is the recruitment of the focal adhesion protein Zyxin to locations of high deformation in actin stress fibers. To demonstrate this behavior, mouse embryonal fibroblasts were transfected with GFP-labelled Zyxin and subjected to external force using the magnetic tweezers. The image sequence displayed in Fig. 5.3 clearly shows a flash of Zyxin within 10 seconds of force application around the area where the bead is bound to the cell and the stress is transmitted to the cytoskeleton.

Fluorescent speckle microscopy

Fluorescent speckle microscopy (FSM) is another technique for fluorescent live cell imaging that is suitable for obtaining quantitative deformation data [Danuser 06]. In this approach, living cells are injected with small amounts of fluorescently labelled monomeric

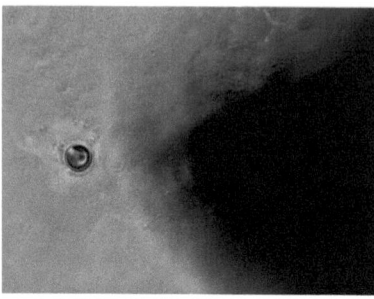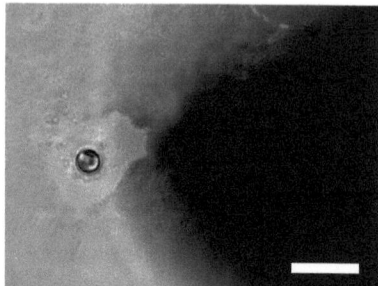

Figure 5.3: Mouse embryonal fibroblast transfected with GFP-Zyxin and a bead bound to the surface. The two images were taken prior to force application (left) and after ten seconds of force application (right). Zyxin is shown in red, a bright field image is overlaid in grayscale. Zyxin is recruited to the area surrounding the bead. Scale bar: 10 μm.

cytoskeletal proteins such as actin or tubulin. These molecules are then incorporated into the cytoskeleton upon polymerization of new filaments. Due to the stochastic distribution of fluorescently labelled monomers, the polymerized actin network appears speckled in the fluorescence microscope (Fig. 5.4a). When these bright spots or speckles are used as strain markers, high-resolution flow maps of the cytoskeleton can be obtained by applying particle tracking algorithms.

In combination with magnetic tweezers microrheology, however, it turned out that FSM only works well in the very flat lamellipodial regions of the cell. While these regions are of interest when studying cell migration and actin polymerization, they are of limited suitability for cell rheology since they only consist of a very thin and stiff actin network layer that is tightly linked to the underlying substrate via the cell membrane. The FSM approach was therefore not pursued further in the course of this work.

5.1.3 Improvements of the software

A general problem with live cell imaging in combination with magnetic tweezers microrheology is that the two methods require different modes of illumination. While bead displacements for microrheology are obtained in bright field transmitted-light illumination, the methods of quantitative live cell imaging described above require fluorescent incident-light or confocal microscopy. In order to combine both methods, the micro-

5.1 Limitations and improvements of the experimental setup

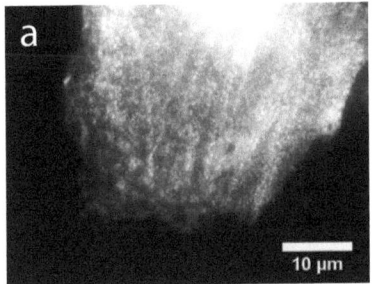 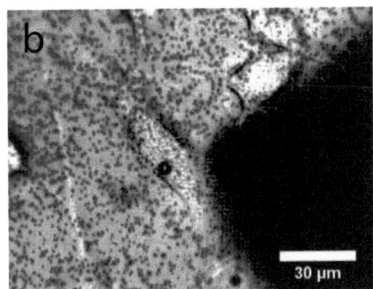

Figure 5.4: a) Example of Fluorescence Speckle Microscopy (FSM) in the lamellipodia of a living Ptk1 cell. The cell was injected with fluorescently labelled G-actin monomers, which are stochastically incorporated into the cytoskeleton upon polymerization. The resulting speckled appearance can be used to track the cytoskeletal flow. b) Example of traction microscopy in combination with magnetic tweezers microrheology. NIH 3T3 fibroblast cells (greyscale bright-field image) were seeded on a soft polyacrylamide (PAA) gel with fluorescent marker beads embedded (magenta). When force is applied to beads bound to the cell surface with magnetic tweezers, the resulting stress transmitted throughout the cell onto the substrate can be determined from the deformation field of the substrate (green arrows).

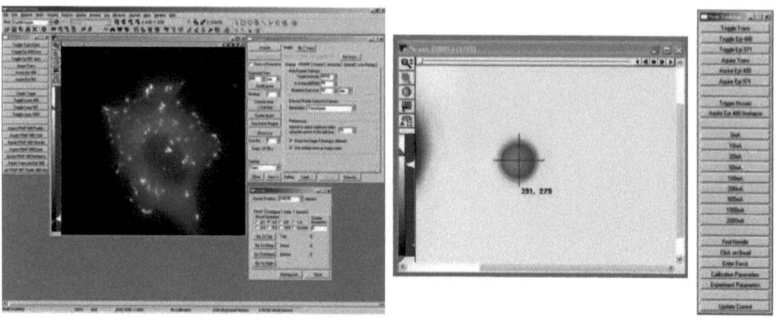

Figure 5.5: Screenshot of an alternative implementation of the magnetic tweezers software using the commercial MetaMorph microscopy software. The time-lapse imaging functionality and user friendliness (left) are expanded by custom functions for bead tracking (middle) and magnetic tweezers control (right) using the built-in scripting interface.

scope has to be switched quickly between both modes of illumination. Moreover, the real-time bead tracking and force control described in this work is done using custom written software, whereas most live cell microscopy setups rely on commercial software packages such as MetaMorph or manufacturer-supplied software. Whereas such software packages provide additional functionality such as automated microscope control, software autofocus, or time lapse- and stack recording, they usually lack the real-time functionality necessary for microrheology.

In principle, however, such commercial microscopy software can be adapted for microrheology experiments if a programming interface for customization is provided. In the case of MetaMorph, a scripting enginge based on Visual Basic is included that provides access to all functions of the software and to the raw image data. In order to evaluate the feasibility of this interface for microrheology experiments, bead and needle tracking algorithms were implemented using this scripting interface, and a graphical menu for controlling the magnet current was added (Fig. 5.5). While this newly added functionality proved to be sufficient for force application to beads during time-lapse live cell experiments, it is not fast enough and lacks the time resolution for conducting microrheology experiments.

A new possbility is to use the μ-manager environment, an open source library for cell biology experiments and live cell imaging. It includes low-level drivers for most common microscope and camera devices and offers both the flexibility of a C++ library and the user-friendliness of commercial software. For user interaction and image processing, a graphical interface using the ImageJ image processing package is included. Alternatively, the full functionality of the library can be accessed from within Matlab. In principle, it is possible to modify the measurement software developed in this work to use the μ-manager drivers and functions instead of the ccd.lib library. The time necessary to implement and test such a new software, however, was beyond the scope of this work and is left as a project for the future.

5.2 Discussion of the experimental results

5.2.1 Power law rheology

In the first part of the experimental results, it was shown that the power law creep modulus $J(t) = J_0(t/t_0)^\beta$ of living cells also holds in the nonlinear regime of large forces. The creep response of living cells can therefore be described by only two parameters: a compliance prefactor J_0 the inverse of which can be interpreted as stiffness, and a power law exponent β that characterizes the time dependence of the creep response between purely elastic ($\beta = 0$) and purely viscous ($\beta = 1$) behavior. The third parameter, t_0, has been set to $t_0 = 1$ s. The choice of t_0 influences the value of J_0, which corresponds to the creep compliance $J(t)$ evaluated at time $t = t_0$.

This raises the question whether the choice of t_0 also influences the force dependence of J_0, or the amount of stress stiffening. The answer is that in the linear regime, J_0 and β are not independent and are linked via the master equation (3.1.11), therefore the choice of t_0 is arbitrary. In the nonlinear step creep measurements, $t_0 = 1$ s corresponds to the timescale of the experiment since each force step lasts for one second. In this case, the prefactor J_0 gives a measure for the displacement after each force step, and hence for the stiffness, independent of the power law exponent. Any other choice of t_0 would lead to a artificial entanglement between J_0 and β. In any case, the two parameters of the power law cannot be completely independent due to the common physical origin of energy storage and dissipation in the cytoskeleton. The conclusion from the experimental results was that the cytoskeletal prestress as mechanical master parameter determines both the elastic (J_0) and dissipative (β) properties of the cell.

5.2.2 Variation of parameters between cells

The stiffness of large numbers of cells always exhibited a log-normal distribution, which corresponds to a normal distribution of power law exponents. This has been observed earlier by other authors using different techniques, and seems to be a generic feature of cell mechanics [Fabry 03, Desprat 05, Balland 06, Hiratsuka 09]. In fact, many properties of biological systems are log-normally distributed between individuals. Log-normal distributions can arise, e.g., from an underlying stochastic process where a variable is the result of multiplication of many independent small positive random numbers. Such stochastic event chains might also be responsible for the log-normal distribution of cell stiffness. Further investigations to elucidate the origin of cell stiffness distributions are certainly worthwhile.

While the shape of the observed probability distributions apparently does not depend on experimental conditions, their width certainly does. For instance, the cell size, which influences the cell mechanical properties, varies widely between individual cells of the same cell type and in the same culture dish. The degree of bead internalization, which cannot be controlled in any other way than by ensuring identical binding times for all beads, also varies considerably from bead to bead. Further effort should be taken to achieve identical measurement conditions for every cell and every bead. An interesting approach has been taken by the Okajima group [Hiratsuka 09] by using comb-like structures for cell culture, leading to identical cell sizes and therefore a considerably smaller variance of mechanical properties between individual cells. Such techniques seem promising for obtaining reproducible conditions in quantitative live cell experiments.

5.2.3 Role of prestress

The experimental results of this work show that not only the linear, but also the nonlinear mechanical properties of cells depend directly on the level of myosin-generated intracellular prestress. This contributes to the picture of prestress as the mechanical key parameters in living cells. The importance of cytoskeletal tension, however, goes far beyond cell rheology: it influences not only the mechanical properties, but also how mechanical signals are transmitted and sensed by cells. Prestress therefore plays a vital role in mechanotransduction and mechanochemical signaling, and all related biological functions. In turn, many biochemical signalling pathways in the cell regulate the degree of myosin light chain phosphorylation and, hence, the level of myosin-generated prestress.

5.2 Discussion of the experimental results

It should be noted that the active cytoskeletal prestress has not been measured directly in the experiments described in this work. The relationship between nonlinear mechanical properties and prestress was based solely on the already established proportionality between linear stiffness and contractile tone of the cell. Direct prestress measurements require a combination of magnetic tweezers microrheology and traction microscopy.

As a proof of concept that the setup described in this work can be combined with traction microscopy, preliminary experiments were carried out. Cells were seeded on a flexible polyacrylamide (PAA) substrate with embedded fluorescent strain markers. Fluorescent images of the gel surface were taken before, during and after force application to a bead bound to the cell surface (Fig. 5.4b), but the bead displacement could not be measured. In order to get bead displacement data and derive mechanical parameters, the measurement software has to be expanded by imaging capabilities as described above (section 5.1.3). In addition, a baseline image of the substrate beneath every measured cell after detaching the cells has to be taken. Because of time, such measurements were not carried out in the present work. This should be done in the near future since direct measurements of prestress will considerably strengthen the conclusions from the experimental results.

5.2.4 Active or passive stress stiffening?

In the nonlinear high force regime, stress stiffening was generally observed: with increasing force, cells appeared increasingly stiffer. This behavior was interpreted in terms of passive stiffening due to a generic nonlinear stress-strain relationship of the cell. It cannot be ruled out completely, however, that this force dependence is actually a time dependence: the stiffening could be a purely active response of the cell over time – e.g., a slowly increasing contractile activation in response to external force.

There are, however, several points against active stiffening but in favor of a passive stiffening mechanism. First of all, stress stiffening is still observed after inhibition of myosin with a non-muscle myosin-IIA inhibitor (blebbistatin), and after biochemical disruption of the actin cytoskeleton by cytochalasin-D. This rules out actomyosin-based contraction as origin of active stiffening. Furthermore, passive stress stiffening and exponential stress-strain relationships are observed in a wide range of biological materials from in-vitro biopolymer networks up to whole tissue strips. A similar behavior would therefore be expected for cells, too. Indeed, it has been shown recently that whole cells permanently crosslinked by glutaraldehyde show a static exponential stress-strain curve [Fernandez 08].

5.2.5 Quantitative analysis of adhesion strength

The percentage of detached beads at a certain force level in step-creep experiments was used as a measure of bead binding strength. Beads that are coated with extracellular matrix (ECM) proteins such as fibronectin appear to the cell as a substrate for adhesion. In consequence, cells bind these beads by the exact same mechanism as they adhere to the ECM. The binding strength of the beads is therefore a measure for the adhesion strength of cells to the ECM and makes it possible to interpret the bead detachment data from a biological perspective.

A more quantitative analysis of the binding strength is, however, desirable. This could be accomplished, e.g., by ramping up the force at a certain rate for every bead until it detaches – a technique called dynamic force spectroscopy [Merkel 99, Prechtel 02]. The characteristic detachment force could then be analyzed statistically and compared to model predictions. The magnetic tweezers setup and measurement software is currently being improved in order to make force spectroscopy experiments possible.

5.2.6 Non-recovery, plasticity, and force reversal

The incomplete recovery of bead displacement after creep and the steady state response after several cycles of creep and recovery have been interpreted as permanent plastic deformation and preconditioning, caused by irreversible force-induced structural changes in the cytoskeleton. It would be interesting to observe the bead displacement after reverting the direction of force application. With the current unipolar magnetic tweezers implementation, this is not possible.

There are a number of possible approaches to overcome this limitation. The most obvious would be to use two opposing magnetic needles in order to apply alternating forces in positive and negative direction. Finite element calculations and experimental tests showed that the magnetic induction in the inactive core distorts the field gradient and reduces the applicable force considerably. A more promising idea is to use a fast rotating stage to rotate the cell by 180°, whereas the magnet remains in place. The magnetic tweezers setup is currently being upgraded with such a rotating stage with automatic offset correction.

5.3 Discussion of the theoretical model

5.3.1 Comparison to the Soft Glassy Rheology (SGR) model

In the theoretical part of this work, a model of cell mechanics was developed that combines elements of soft glassy rheology, the wormlike chain model of semiflexible polymers, and the sliding filament model of muscle contraction. It demonstrates how power law rheology, nonlinear elasticity and force generation in biological materials can arise from generic structural properties independent of the actual molecular constituents of the system. All viscous dissipative properties in this generalized sliding filament (GSF) model of cell mechanics are due to microscopic unbinding of weak biological bonds.

The main contribution of the SGR model is the concept that an exponential distribution of yield energies together with the exponential dependence of element lifetimes on the activation energy leads to a power law dependence of the linear viscoelastic moduli on time or frequency. The power-law regime observed in cell mechanics is limited to a certain time regime: at times shorter than a few milliseconds, cell rheology is dominated by thermal fluctuations of cytoskeletal filaments, whereas at times larger than 10-20 seconds, active processes and reorganization within the cytoskeleton prevent the derivation of classical rheological parameters. Therefore, it is not necessary to postulate a true analytic exponential distribution of energies – a broad distribution of element energies is sufficient. The analytical proof of the existence of a glass transition in the SGR model, which also depends on a true exponential energy distribution, is not relevant for the description of living cells.

The central parameter of the SGR model, the noise temperature x, does not appear directly in the model developed in this work. Instead, the real thermodynamic temperature is used. Implicitly, x appears in the model as the ratio between the typical element energy and the thermal energy, $k_B T$. Compared to colloids or foams for which the original SGR model was intended, the weak interactions in biological systems dissociate spontaneously due to thermal activation, and $x > 1$ in all cases. The introduction of an effective noise temperature to describe activated hopping of elements between traps is therefore not necessary.

Another difference to the SGR model is that the elements have nonlinear force-extension curves, in contrast to the Hookean spring-like behavior of the SGR elements. Since the yield force depends on relative extension only and is identical for all elements, the stored elastic energy $E = \int f dl$ is proportional to the displacement l, and not to l^2 as in the SGR model.

For simplicity, the yield force is assumed to be constant for all elements, only bond lengths and hence yield energies are distributed. In reality, though, the strength of bonds is probably also distributed since many different bond types are present in the system.

5.3.2 Relation to the „Glassy Wormlike Chain" (GWLC)

The "Glassy Wormlike Chain" (GWLC) model first introduced in [Semmrich 07] combines the Wormlike Chain (WLC) description of semiflexible polymers with elements of Soft Glassy Rheology (SGR). In this model, the dissipative properties arise primarily from the friction of fluctuating filaments against the viscous background fluid, as in the classical wormlike chain picture [Gittes 98]. Sticky interactions with crosslinks or other filaments slow down the relaxation of the filament fluctuations, therefore accounting for long-lasting, power-law like relaxation dynamics.

Although the concept may seem similar to the model described in this work, the two approaches are fundamentally different. Whereas the GWLC model starts from a *fast* relaxation time spectrum that is *slowed down* by sticky interactions, the model described here starts from purely *static* elastic behavior that exhibits stress relaxation *only* due to unbinding of sticky interactions. Eventually, the phenomenological behavior predicted by both models for intermediate timescales is similar.

The GWLC model accurately and quantitatively describes the fluctuation spectrum of entangled actin networks over many orders of magnitude of frequency [Semmrich 07]. When the model is applied to cells, however, the resulting values for the background viscosity are two orders of magnitude larger than for pure entangled actin networks. One possible explanation is that the crowded environment of the cytoplasm leads to a higher effective friction coefficient. It seems more likely that thermal fluctuations are mostly stretched out in prestressed networks such as the cytoskeleton and do not contribute significantly to the time dependence observed in typical cell rheology experiments. On the one hand, GWLC might not be the model of choice to describe cell rheology in the intermediate time regime of milliseconds to seconds. The advantage of the GWLC approach, on the other hand, is that it describes the transistion of cell rheology to the high-frequency regime above 1 kHz or below 1 ms where thermal filament fluctuations become relevant.

5.3.3 Limitations of the numerical results

The stress-strain relationship of individual elements and the steady-state energy distribution of the model were derived analytically, but the actual time-and force-dependent behavior was only investigated by numerical Monte Carlo studies. Such simulations have the character of experiments, and their explanatory power is limited. It would be desirable to have analytical expressions for the observed variables such as the creep response $J(t)$ in order to be able to fit the parameters of the model to experimental data. This will probably require several simplifications of the model and was not pursued further in the course of this work.

With the Kinetic Monte Carlo (KMC) algorithm used in this work, it is not possible to simulate systems where reaction rates change in between two events. This is the case for experimental protocols where the applied force or strain is a continous function of time, such as a force ramp. Various extensions to the KMC algorithm exist that give accurate results for time-dependent rates. Since they are much more complex and numerically expensive, their implementation was left as a possible future extension.

5.3.4 Uniaxial force generation

The geometry of the model was intended to describe microrheology experiments on cells. In this case, the prestress of the cytoskeleton is isotropic from the point of view of the bead, and the bead remains at its position as long as no external force is applied. Displacement of the bead across the cell surface leads to shear of the underlying cytoskeleton that is randomly oriented with respect to the prestress force on the individual elements. This is geometrically analogous to the situation in a shear rheometer where rotation of one of the plates is the only degree of freedom.

The geometry of experiments where external force and internal prestress act in parallel is fundamentally different. Examples for such a situation are uniaxial stretching of contractile cells or networks, or measurements on isolated stress fibers or contractile smooth muscle tissue. In order to describe uniaxial force generation, the model has to be modified by breaking the symmetry of force generation and using Huxley-like attachment rules where elements can only attach on one side of the force-generating regime.

5.3.5 Applicability to other systems

The rheological response of reconstituted cytoskeletal networks resembles that of living cells [Koenderink 09]: power law rheology and stress stiffening with prestress-dependent parameters. Many other biopolymer systems such as reconstituted collagen networks, smooth muscle strips or spider silk exhibit similar behavior. The model explains these observations in a mechanistic way: stress stiffening is due to geometrical alignment and nonlinear elasticity of elements, and power law stress relaxation and creep arise from microscopic yielding and broadly distributed lifetimes of elements. It seems obvious that these mechanisms are not limited to the cytoskeleton, but are a universal property of many biological materials.

Appendix

Table of simulation parameters

Name	Description	Value	Unit	Reference
T	temperature	310	K	physiological temperature
k_B	Boltzmann constant	$1.38 \cdot 10^{-23}$	J/K	relates energy to T
ℓ_p	persistence length of elements	17	μm	F-actin, [Gardel 04]
γ_c	maximum strain of elements	1.5	%	
f_c	critical force	~ 4	pN	typical bond strength
c	ratio of catch/slip bond length	3		actin-myosin, [Guo 06]
d	ratio of catch/slip yield force	0.2		actin-myosin, [Guo 06]
N_e	number of elements	15000		
τ_0	intrinsic time constant	0.01	s	

Bibliography

[Ashkin 70] A. Ashkin. *Acceleration and Trapping of Particles by Radiation Pressure.* Phys. Rev. Lett., vol. 24, no. 4, pages 156–159, Jan 1970.

[Balland 06] Martial Balland, Nicolas Desprat, Delphine Icard, Sophie Fereol, Atef Asnacios, Julien Browaeys, Sylvie Henon & Francois Gallet. *Power laws in microrheology experiments on living cells: Comparative analysis and modeling.* Phys Rev E Stat Nonlin Soft Matter Phys, vol. 74, no. 2 Pt 1, page 021911, Aug 2006.

[Bausch 98] A. R. Bausch, F. Ziemann, A. A. Boulbitch, K. Jacobson & E. Sackmann. *Local measurements of viscoelastic parameters of adherent cell surfaces by magnetic bead microrheometry.* Biophys J, vol. 75, no. 4, pages 2038–2049, Oct 1998.

[Bausch 06] A. R. Bausch & K. Kroy. *A bottom-up approach to cell mechanics.* Nat Phys, vol. 2, no. 4, pages 231–238, April 2006.

[Bell 78] G. I. Bell. *Models for the specific adhesion of cells to cells.* Science, vol. 200, no. 4342, pages 618–627, May 1978.

[Binnig 86] G. Binnig, C. F. Quate & Ch. Gerber. *Atomic Force Microscope.* Phys. Rev. Lett., vol. 56, no. 9, pages 930–933, Mar 1986.

[Bouchaud 92] J. P. Bouchaud. *Weak ergodicity breaking and aging in disordered systems.* J. Phys. I France, vol. 2, no. 9, pages 1705–1713, sep 1992.

[Brent 73] Richard P. Brent. *Algorithms for minimisation without derivatives.* Prentice Hall, 1973.

[Bursac 05] Predrag Bursac, Guillaume Lenormand, Ben Fabry, Madavi Oliver, David A Weitz, Virgile Viasnoff, James P Butler & Jeffrey J Fredberg. *Cytoskeletal remodelling and slow dynamics in the living cell.* Nat Mater, vol. 4, no. 7, pages 557–561, Jul 2005.

[Bustamante 94] C. Bustamante, J. F. Marko, E. D. Siggia & S. Smith. *Entropic elasticity of lambda-phage DNA.* Science, vol. 265, no. 5178, pages 1599–1600, Sep 1994.

[Crick 50] F. H. C. Crick & A. F. W. Hughes. *The Physical Properties of Cytoplasm: A Study by Means of the Magnetic Particle Method.* Exp. Cell Res., vol. 1, pages 37–80, 1950.

[Danuser 06] Gaudenz Danuser & Clare M Waterman-Storer. *Quantitative fluorescent speckle microscopy of cytoskeleton dynamics.* Annu Rev Biophys Biomol Struct, vol. 35, pages 361–387, 2006.

[Deng 06] Linhong Deng, Xavier Trepat, James P Butler, Emil Millet, Kathleen G Morgan, David A Weitz & Jeffrey J Fredberg. *Fast and slow dynamics of the cytoskeleton.* Nat Mater, vol. 5, no. 8, pages 636–640, Aug 2006.

[Desprat 05] Nicolas Desprat, Alain Richert, Jacqueline Simeon & Atef Asnacios. *Creep function of a single living cell.* Biophys J, vol. 88, no. 3, pages 2224–2233, Mar 2005.

[DiDonna 06] B. A. DiDonna & Alex J Levine. *Filamin cross-linked semiflexible networks: fragility under strain.* Phys Rev Lett, vol. 97, no. 6, page 068104, Aug 2006.

[Edelstein-Keshet 98] L. Edelstein-Keshet & G. B. Ermentrout. *Models for the length distributions of actin filaments: I. Simple polymerization and fragmentation.* Bull Math Biol, vol. 60, no. 3, pages 449–475, May 1998.

[Erdmann 04] T. Erdmann & U. S. Schwarz. *Stability of adhesion clusters under constant force.* Phys Rev Lett, vol. 92, no. 10, page 108102, Mar 2004.

[Evans 97] E. Evans & K. Ritchie. *Dynamic strength of molecular adhesion bonds.* Biophys J, vol. 72, no. 4, pages 1541–1555, Apr 1997.

[Fabry 01]	B. Fabry, G. N. Maksym, J. P. Butler, M. Glogauer, D. Navajas & J. J. Fredberg. *Scaling the microrheology of living cells.* Phys Rev Lett, vol. 87, no. 14, page 148102, Oct 2001.
[Fabry 03]	Ben Fabry & Jeffrey J Fredberg. *Remodeling of the airway smooth muscle cell: are we built of glass?* Respir Physiol Neurobiol, vol. 137, no. 2-3, pages 109–124, Sep 2003.
[Feneberg 04]	Wolfgang Feneberg, Martin Aepfelbacher & Erich Sackmann. *Microviscoelasticity of the apical cell surface of human umbilical vein endothelial cells (HUVEC) within confluent monolayers.* Biophys J, vol. 87, no. 2, pages 1338–1350, Aug 2004.
[Fernandez 06]	Pablo Fernandez, Pramod A Pullarkat & Albrecht Ott. *A master relation defines the nonlinear viscoelasticity of single fibroblasts.* Biophys J, vol. 90, no. 10, pages 3796–3805, May 2006.
[Fernandez 07]	Pablo Fernandez, Lutz Heymann, Albrecht Ott, Nuri Aksel & Pramod A Pullarkat. *Shear rheology of a cell monolayer.* New Journal of Physics, vol. 9, no. 11, page 419, 2007.
[Fernandez 08]	Pablo Fernandez & Albrecht Ott. *Single cell mechanics: stress stiffening and kinematic hardening.* Phys Rev Lett, vol. 100, no. 23, page 238102, Jun 2008.
[Fredberg 89]	J. J. Fredberg & D. Stamenovic. *On the imperfect elasticity of lung tissue.* J Appl Physiol, vol. 67, no. 6, pages 2408–2419, Dec 1989.
[Fredberg 96]	J. J. Fredberg, K. A. Jones, M. Nathan, S. Raboudi, Y. S. Prakash, S. A. Shore, J. P. Butler & G. C. Sieck. *Friction in airway smooth muscle: mechanism, latch, and implications in asthma.* J Appl Physiol, vol. 81, no. 6, pages 2703–2712, Dec 1996.
[Freundlich 23]	H. Freundlich & W. Seifriz. *Über die Elastizität von Solen und Gelen.* Zeitsch. phys. Chem., vol. 104, pages 233–261, 1923.
[Fung 93]	Y. C. Fung. Mechanical properties of living tissues. Springer-Verlag New York, Inc., 1993.
[Gardel 04]	M. L. Gardel, J. H. Shin, F. C. MacKintosh, L. Mahadevan, P. Matsudaira & D. A. Weitz. *Elastic behavior of cross-linked and bundled*

actin networks. Science, vol. 304, no. 5675, pages 1301–1305, May 2004.

[Gardel 06a]   M. L. Gardel, F. Nakamura, J. Hartwig, J. C. Crocker, T. P. Stossel & D. A. Weitz. *Stress-dependent elasticity of composite actin networks as a model for cell behavior.* Phys Rev Lett, vol. 96, no. 8, page 088102, Mar 2006.

[Gardel 06b]   M. L. Gardel, F. Nakamura, J. H. Hartwig, J. C. Crocker, T. P. Stossel & D. A. Weitz. *Prestressed F-actin networks cross-linked by hinged filamins replicate mechanical properties of cells.* Proc Natl Acad Sci U S A, vol. 103, no. 6, pages 1762–1767, Feb 2006.

[Gillespie 77]   Daniel T. Gillespie. *Exact stochastic simulation of coupled chemical reactions.* The Journal of Physical Chemistry, vol. 81, no. 25, pages 2340–2361, December 1977.

[Gittes 98]   F. Gittes & F. C. MacKintosh. *Dynamic shear modulus of a semiflexible polymer network.* Phys. Rev. E, vol. 58, no. 2, pages R1241–R1244, Aug 1998.

[Gosse 02]   Charlie Gosse & Vincent Croquette. *Magnetic tweezers: micromanipulation and force measurement at the molecular level.* Biophys J, vol. 82, no. 6, pages 3314–3329, Jun 2002.

[Guck 01]   J. Guck, R. Ananthakrishnan, H. Mahmood, T. J. Moon, C. C. Cunningham & J. Kas. *The optical stretcher: a novel laser tool to micromanipulate cells.* Biophys J, vol. 81, no. 2, pages 767–784, Aug 2001.

[Guilford 92]   W. H. Guilford & R. W. Gore. *A novel remote-sensing isometric force transducer for micromechanics studies.* Am J Physiol, vol. 263, no. 3 Pt 1, pages C700–C707, Sep 1992.

[Guo 06]   Bin Guo & William H Guilford. *Mechanics of actomyosin bonds in different nucleotide states are tuned to muscle contraction.* Proc Natl Acad Sci U S A, vol. 103, no. 26, pages 9844–9849, Jun 2006.

[Hai 88]	C. M. Hai & R. A. Murphy. *Regulation of shortening velocity by cross-bridge phosphorylation in smooth muscle*. Am J Physiol, vol. 255, no. 1 Pt 1, pages C86–C94, Jul 1988.
[Heilbronn 22]	A. Heilbronn. *Eine neue Methode zur Bestimmung der Viskosität lebender Protoplasten*. Jahrbuch für wiss. Botanik, vol. 61, pages 284–338, 1922.
[Heussinger 07]	Claus Heussinger, Boris Schaefer & Erwin Frey. *Nonaffine rubber elasticity for stiff polymer networks*. Phys Rev E Stat Nonlin Soft Matter Phys, vol. 76, no. 3 Pt 1, page 031906, Sep 2007.
[Hiramoto 69]	Y. Hiramoto. *Mechanical properties of the protoplasm of the sea urchin egg. I. Unfertilized egg*. Exp Cell Res, vol. 56, no. 2, pages 201–208, Aug 1969.
[Hiratsuka 09]	Shinichiro Hiratsuka, Yusuke Mizutani, Masahiro Tsuchiya, Koichi Kawahara, Hiroshi Tokumoto & Takaharu Okajima. *The number distribution of complex shear modulus of single cells measured by atomic force microscopy*. Ultramicroscopy, Mar 2009.
[Hoffman 06]	Brenton D Hoffman, Gladys Massiera, Kathleen M Van Citters & John C Crocker. *The consensus mechanics of cultured mammalian cells*. Proc Natl Acad Sci U S A, vol. 103, no. 27, pages 10259–10264, Jul 2006.
[Hoffman 07]	Brenton D Hoffman, Gladys Massiera & John C Crocker. *Fragility and mechanosensing in a thermalized cytoskeleton model with forced protein unfolding*. Phys Rev E Stat Nonlin Soft Matter Phys, vol. 76, no. 5 Pt 1, page 051906, Nov 2007.
[Hu 03]	Shaohua Hu, Jianxin Chen, Ben Fabry, Yasushi Numaguchi, Andrew Gouldstone, Donald E Ingber, Jeffrey J Fredberg, James P Butler & Ning Wang. *Intracellular stress tomography reveals stress focusing and structural anisotropy in cytoskeleton of living cells*. Am J Physiol Cell Physiol, vol. 285, no. 5, pages C1082–C1090, Nov 2003.
[Huxley 57]	A. F. Huxley. *Muscle structure and theories of contraction*. Prog Biophys Biophys Chem, vol. 7, pages 255–318, 1957.

[Ingber 03]   Donald E Ingber. *Tensegrity I. Cell structure and hierarchical systems biology.* J Cell Sci, vol. 116, no. Pt 7, pages 1157–1173, Apr 2003.

[Kasza 09]   K. E. Kasza, F. Nakamura, S. Hu, P. Kollmannsberger, N. Bonakdar, B. Fabry, T. P. Stossel, N. Wang & D. A. Weitz. *Filamin A is essential for active cell stiffening but not passive stiffening under external force.* Biophys J, vol. 96, no. 10, pages 4326–4335, May 2009.

[Koenderink 09] Gijsje H Koenderink, Zvonimir Dogic, Fumihiko Nakamura, Poul M Bendix, Frederick C MacKintosh, John H Hartwig, Thomas P Stossel & David A Weitz. *An active biopolymer network controlled by molecular motors.* Proc Natl Acad Sci U S A, vol. 106, no. 36, pages 15192–15197, Sep 2009.

[Kong 09]   Fang Kong, Andres J Garcia, A. Paul Mould, Martin J Humphries & Cheng Zhu. *Demonstration of catch bonds between an integrin and its ligand.* J Cell Biol, vol. 185, no. 7, pages 1275–1284, Jun 2009.

[Laurent 02]   V. M. Laurent, S. Henon, E. Planus, R. Fodil, M. Balland, D. Isabey & F. Gallet. *Assessment of mechanical properties of adherent living cells by bead micromanipulation: comparison of magnetic twisting cytometry vs optical tweezers.* J Biomech Eng, vol. 124, no. 4, pages 408–421, Aug 2002.

[Lenormand 04] G. Lenormand, E. Millet, B. Fabry, J. P. Butler & J. J. Fredberg. *Linearity and time-scale invariance of the creep function in living cells.* J R Soc Interface, vol. 1, no. 1, pages 91–97, Nov 2004.

[MacKintosh 95] MacKintosh, Käs & Janmey. *Elasticity of semiflexible biopolymer networks.* Phys Rev Lett, vol. 75, no. 24, pages 4425–4428, Dec 1995.

[Matthews 04]   Benjamin D. Matthews, David A. LaVan, Darryl R. Overby, John Karavitis & Donald E. Ingber. *Electromagnetic needles with submicron pole tip radii for nanomanipulation of biomolecules and living cells.* Applied Physics Letters, vol. 85, no. 14, pages 2968–2970, 2004.

Bibliography

[Maxwell 67] J.C. Maxwell. *On the Dynamical Theory of Gases*. Philos. Trans. R. Soc. London, vol. 157, pages 49–88, 1867.

[McMahon 84] T. A. McMahon. *Muscles, reflexes, and locomotion*. Princeton University Press, Princeton University Press, 41 William Street, Princeton, New Jersey, 1984.

[Medalia 02] Ohad Medalia, Igor Weber, Achilleas S Frangakis, Daniela Nicastro, Gunther Gerisch & Wolfgang Baumeister. *Macromolecular architecture in eukaryotic cells visualized by cryoelectron tomography*. Science, vol. 298, no. 5596, pages 1209–1213, Nov 2002.

[Merkel 99] R. Merkel, P. Nassoy, A. Leung, K. Ritchie & E. Evans. *Energy landscapes of receptor-ligand bonds explored with dynamic force spectroscopy*. Nature, vol. 397, no. 6714, pages 50–53, Jan 1999.

[Mierke 08a] Claudia Tanja Mierke, Philip Kollmannsberger, Daniel Paranhos Zitterbart, James Smith, Ben Fabry & Wolfgang Heinrich Goldmann. *Mechano-coupling and regulation of contractility by the vinculin tail domain*. Biophys J, vol. 94, no. 2, pages 661–670, Jan 2008.

[Mierke 08b] Claudia Tanja Mierke, Daniel Paranhos Zitterbart, Philip Kollmannsberger, Carina Raupach, Ursula Schlötzer-Schrehardt, Tamme Weyert Goecke, Jürgen Behrens & Ben Fabry. *Breakdown of the endothelial barrier function in tumor cell transmigration*. Biophys J, vol. 94, no. 7, pages 2832–2846, Apr 2008.

[Mijailovich 00] S. M. Mijailovich, J. P. Butler & J. J. Fredberg. *Perturbed equilibria of myosin binding in airway smooth muscle: bond-length distributions, mechanics, and ATP metabolism*. Biophys J, vol. 79, no. 5, pages 2667–2681, Nov 2000.

[Onck 05] P. R. Onck, T. Koeman, T. van Dillen & E. van der Giessen. *Alternative explanation of stiffening in cross-linked semiflexible networks*. Phys Rev Lett, vol. 95, no. 17, page 178102, Oct 2005.

[Pauli 05] J. Pauli. *Development of a class library for controlling a CCD camera in a complex, biological environment*. Seminar talk, Universität Erlangen, Jan 24, 2005.

[Pereverzev 05]   Yuriy V Pereverzev, Oleg V Prezhdo, Manu Forero, Evgeni V Sokurenko & Wendy E Thomas. *The two-pathway model for the catch-slip transition in biological adhesion.* Biophys J, vol. 89, no. 3, pages 1446–1454, Sep 2005.

[Prechtel 02]   K. Prechtel, A. R. Bausch, V. Marchi-Artzner, M. Kantlehner, H. Kessler & R. Merkel. *Dynamic force spectroscopy to probe adhesion strength of living cells.* Phys Rev Lett, vol. 89, no. 2, page 028101, Jul 2002.

[Radmacher 92]   M. Radmacher, R. W. Tillamnn, M. Fritz & H. E. Gaub. *From molecules to cells: imaging soft samples with the atomic force microscope.* Science, vol. 257, no. 5078, pages 1900–1905, Sep 1992.

[Raupach 07]   Carina Raupach, Daniel Paranhos Zitterbart, Claudia T Mierke, Claus Metzner, Frank A Müller & Ben Fabry. *Stress fluctuations and motion of cytoskeletal-bound markers.* Phys Rev E Stat Nonlin Soft Matter Phys, vol. 76, no. 1 Pt 1, page 011918, Jul 2007.

[Rondelez 05]   Yannick Rondelez, Guillaume Tresset, Takako Nakashima, Yasuyuki Kato-Yamada, Hiroyuki Fujita, Shoji Takeuchi & Hiroyuki Noji. *Highly coupled ATP synthesis by F1-ATPase single molecules.* Nature, vol. 433, no. 7027, pages 773–777, Feb 2005.

[Schmidt 98]   Anja Schmidt & Michael N. Hall. *Signaling to the actin cytoskeleton.* Annual Review of Cell and Developmental Biology, vol. 14, no. 1, pages 305–338, 1998.

[Seifert 00]   U. Seifert. *Rupture of multiple parallel molecular bonds under dynamic loading.* Phys Rev Lett, vol. 84, no. 12, pages 2750–2753, Mar 2000.

[Semmrich 07]   Christine Semmrich, Tobias Storz, Jens Glaser, Rudolf Merkel, Andreas R Bausch & Klaus Kroy. *Glass transition and rheological redundancy in F-actin solutions.* Proc Natl Acad Sci U S A, vol. 104, no. 51, pages 20199–20203, Dec 2007.

[Shenoy 02]   Aroon Shenoy. *Estimating the Unrecovered Strain During a Creep Recovery Test from the Material's Volumetric-flow Rate.* International Journal of Pavement Engineering, vol. 3, no. 1, pages 29–34, 2002.

Bibliography

[Simson 98] D. A. Simson, F. Ziemann, M. Strigl & R. Merkel. *Micropipet-based pico force transducer: in depth analysis and experimental verification*. Biophys J, vol. 74, no. 4, pages 2080–2088, Apr 1998.

[Sollich 97] Peter Sollich, Fran çois Lequeux, Pascal Hébraud & Michael E. Cates. *Rheology of Soft Glassy Materials*. Phys. Rev. Lett., vol. 78, no. 10, pages 2020–2023, Mar 1997.

[Sollich 98] Peter Sollich. *Rheological constitutive equation for a model of soft glassy materials*. Phys. Rev. E, vol. 58, no. 1, pages 738–759, Jul 1998.

[Stamenović 04] Dimitrije Stamenović, Bela Suki, Ben Fabry, Ning Wang & Jeffrey J Fredberg. *Rheology of airway smooth muscle cells is associated with cytoskeletal contractile stress*. J Appl Physiol, vol. 96, no. 5, pages 1600–1605, May 2004.

[Storm 05] Cornelis Storm, Jennifer J Pastore, F. C. MacKintosh, T. C. Lubensky & Paul A Janmey. *Nonlinear elasticity in biological gels*. Nature, vol. 435, no. 7039, pages 191–194, May 2005.

[Thoumine 97] O. Thoumine & A. Ott. *Time scale dependent viscoelastic and contractile regimes in fibroblasts probed by microplate manipulation*. J Cell Sci, vol. 110 (Pt 17), pages 2109–2116, Sep 1997.

[Trepat 07] Xavier Trepat, Linhong Deng, Steven S An, Daniel Navajas, Daniel J Tschumperlin, William T Gerthoffer, James P Butler & Jeffrey J Fredberg. *Universal physical responses to stretch in the living cell*. Nature, vol. 447, no. 7144, pages 592–595, May 2007.

[Trepat 08] Xavier Trepat, Guillaume Lenormand & Jeffrey J. Fredberg. *Universality in cell mechanics*. Soft Matter, vol. 4, pages 1750–1759, 2008.

[Valberg 87] P. A. Valberg & H. A. Feldman. *Magnetic particle motions within living cells. Measurement of cytoplasmic viscosity and motile activity*. Biophys J, vol. 52, no. 4, pages 551–561, Oct 1987.

[Wang 01] N. Wang, K. Naruse, D. Stamenović, J. J. Fredberg, S. M. Mijailovich, I. M. Tolić-Norrelykke, T. Polte, R. Mannix & D. E. Ingber. *Mechanical behavior in living cells consistent with the tensegrity*

	model. Proc Natl Acad Sci U S A, vol. 98, no. 14, pages 7765–7770, Jul 2001.
[Weber 35]	Wilhelm Weber. *Ueber die Elasticitaet der Seidenfaeden.* Ann. Phys. Chem., vol. 34, pages 247–257, 1835.
[Weber 41]	Wilhelm Weber. *Ueber die Elasticitaet fester Koerper.* Ann. Phys. Chem., vol. 54, page 1, 1841.
[Wertheim 47]	M. G. Wertheim. *Memoire sur l'elasticite et la cohesion des principaux tissus du corps humain.* Annales Chimie et de Physique, vol. 21, pages 385–414, 1847.
[Xu 99]	J. Xu, J. F. Casella & T. D. Pollard. *Effect of capping protein, CapZ, on the length of actin filaments and mechanical properties of actin filament networks.* Cell Motil Cytoskeleton, vol. 42, no. 1, pages 73–81, 1999.
[Yagi 61]	K. Yagi. *The mechanical and colloidal properties of Amoeba protoplasm and their relations to the mechanism of amoeboid movement.* Comp Biochem Physiol, vol. 3, pages 73–91, Aug 1961.
[Yamada 00]	S. Yamada, D. Wirtz & S. C. Kuo. *Mechanics of living cells measured by laser tracking microrheology.* Biophys J, vol. 78, no. 4, pages 1736–1747, Apr 2000.
[Ziemann 94]	F. Ziemann, J. Rädler & E. Sackmann. *Local measurements of viscoelastic moduli of entangled actin networks using an oscillating magnetic bead micro-rheometer.* Biophys J, vol. 66, no. 6, pages 2210–2216, Jun 1994.

I want morebooks!

Buy your books fast and straightforward online - at one of the world's fastest growing online book stores! Environmentally sound due to Print-on-Demand technologies.

Buy your books online at
www.get-morebooks.com

Kaufen Sie Ihre Bücher schnell und unkompliziert online – auf einer der am schnellsten wachsenden Buchhandelsplattformen weltweit! Dank Print-On-Demand umwelt- und ressourcenschonend produziert.

Bücher schneller online kaufen
www.morebooks.de

OmniScriptum Marketing DEU GmbH
Heinrich-Böcking-Str. 6-8
D - 66121 Saarbrücken
Telefax: +49 681 93 81 567-9

info@omniscriptum.com
www.omniscriptum.com

Printed by Books on Demand GmbH, Norderstedt / Germany